Preface

RACING successes, of course, helped to bring the Honda name before the public. What made it a household name was its introduction into the U.K. in late 1962 of the 50 c.c. range of motor-cycles and scooterettes. In Britain, this was marked by a non-stop seven-day run by three of these tiny machines, at Goodwood, the Sussex motor-racing circuit. Officially observed by the Auto-Cycle Union (the controlling body of motor-cycle sport in Britain) a C.110D motor-cycle covered 5,897 miles; a C.102 covered 4,935 miles; and a C.100 covered 5,023 miles. The best fuel consumption for the distance bettered 140 m.p.g. and worst 124 m.p.g. The motor-cycle had averaged 35 m.p.h. and the scooterettes only slightly under 30 m.p.h. This won Honda the Maudes Trophy.

Round about this time I was myself conducting a 2,200 miles Press test of a C.100, with permission from the U.K. Branch of European Honda to break it if I could. Despite the hardest driving I could devise, and despite deliberate neglect of even such essentials as oil top-ups and battery checks, the machine not only failed to blow up but also beat all my personal point-to-point and hill-climbing performances for a "50" on test routes I had used for about 200 various models for several magazines.

Such treatment—though good for a story—is not one which I would recommend to the private owner. Hence this book, in the preparation of which I have been admirably assisted by the enthusiastic staff of Honda (U.K.). It is designed to enable the newcomer to the Honda camp, and to motor-cycling, to service his own machine and to keep it in a tip-top running order. For his benefit, I have included much basic data—though this book is designed to supplement, not supplant, the Owner's Handbook supplied by Honda. But the old hand, like myself, has not been forgotten—the specialized chapters contain enough information to permit an experienced motor-cyclist to strip and rebuild a machine if need be.

All the 50 c.c. Honda machines marketed from their introduction to the British market are covered in this book. They are the models C.100, C.102, C.110, C.114, P.50, PC.50, PF.50 and C.50. In addition, much of the data applicable to the C.100 range holds good for the "Monkey Bikes" and the CE.105H Trail Bike. The 90 c.c. machines are the subject of a separate book.—J.T.

Contents

	Preface	iii
1	The Honda 50s	1
2	Handling Hondas	5
3	Basic principles	11
4	Tools	24
5	Be your own doctor	28
6	Pinpointing troubles	35
7	Methodical maintenance	42
8	Fettling your Honda	46
9	Engine overhauls, pushrod o.h.v. models	61
10	Engine overhauls, o.h.c. models	73
11	Suspension and brakes	87
	Appendixes:	93
	A Facts and figures	94
	B Carburettors	96
	C Wiring Diagrams	100
	Index	107

The Pitman Motor-cyclists' Library

The Book of the

HONDA 50

A Practical Guide to the Handling and Maintenance of all Honda 50s up to 1973

John Thorpe

Pitman Publishing

Third Edition 1973

SIR ISAAC PITMAN AND SONS LTD
Pitman House, Parker Street, Kingsway, London, WC2B 5PB
P.O. Box 46038, Portal Street, Nairobi, Kenya

SIR ISAAC PITMAN (AUST.) PTY LTD
Pitman House, 158 Bouverie Street, Carlton, Victoria 3053, Australia

PITMAN PUBLISHING COMPANY S.A. LTD
P.O. Box 11231, Johannesburg, S. Africa

PITMAN PUBLISHING CORPORATION
6 East 43rd Street, New York, N.Y. 10017, U.S.A.

SIR ISAAC PITMAN (CANADA) LTD
495 Wellington Street West, Toronto 135, Canada

THE COPP CLARK PUBLISHING COMPANY
517 Wellington Street West, Toronto 135, Canada

ISBN: 0 273 00229 5

Text set in 8/9 pt. Monotype Times New Roman, printed by letterpress, and bound in Great Britain at The Pitman Press, Bath

(G.4351 : 19)

1 The Honda 50s

SEVEN versions of the ubiquitous 50 c.c. Honda are covered in this Book. They range from the original C.100 that, on its introduction, so effectively outclassed every other machine of its capacity on the market, to the current C.50.

The "first generation" is represented by the two overhead-valve push-rod scooterettes—the C.100 and C.102—and the two tiny motor-cycles, the C.110 and C.114. All shared a variation of the slightly oversquare push-rod engine, but with automatic clutches on the scooterettes and manual clutches and four-speed gearboxes on the motor-cycles. The push-rod engine was also used, of course, in the "Monkey Bike," and the details given for the C.100 are applicable to that model also.

Fig. 1. The Honda C.100/C.102. The machine that began it all—the o.h.v. 50 c.c. C.100. The C.102 was identical save for an electric starter

Completing the range of scooterettes, the C.50 has the latest chain-driven o.h.c. engine, still with the automatic clutch and three-speed gearbox. This is—by Honda standards—a long-stroke unit, with a 39 mm bore and 41·4 mm stroke and it revs at 10,000 r.p.m., even higher than the push-rod engines which themselves turned over at a by no means low rate of 9,500.

Fig. 2. The Honda C.110/C.114. There were only detail differences between the two 50 c.c. o.h.v. motor-cycles

Fig. 3. The Honda P.50. The single-speed "engine with the wheel" moped with 50 c.c. overhead camshaft power unit

By comparison, the moped engines—overhead camshaft jobs in both the P.50 and PC.50 models—are low-revving, although their 5,750 r.p.m. would have been considered unusual, to say the least, only a few years

Fig. 4. The Honda PC.50. Still a single-speeder with o.h.c. 50 c.c. engine this also incorporates rear suspension. Later PC.50s have the PF.50's push-rod engine unit.

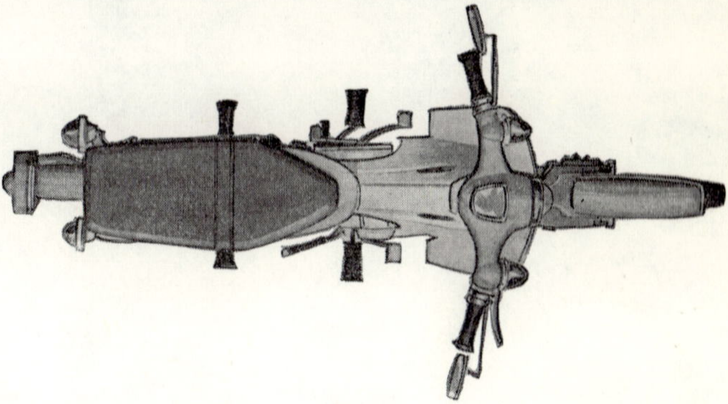

Fig. 5. The Honda C.50. Successor to the C.100, the latest Honda scooterettes have a more powerful 50 c.c. overhead camshaft engine

ago. These are utility machines with that quality of finish that has become a Honda trademark, and with the same ingenuity of design and soundness of construction that characterizes Hondas of all types. Like the rest of the range (apart from the motor-cycles) they have automatic clutches, but only single-speed transmissions.

Frame Numbers. No manufacturer stands still. Improvements are constantly being made as production proceeds—and Honda, with their impressive industrial capacity—are inveterate improvers. Consequently, two machines of apparently identical age and type can have significant differences in specification. The key to this is the frame number.

When ordering spares—*whether for the engine or frame*—it is essential to quote the frame number so that the stockist can identify the machine. The letter prefix to the number is particularly important—e.g., C.50 EO30831.

2 Handling Hondas

THERE is an odd tradition that nobody needs to be taught how to handle a motor-cycle or scooterette. Newcomers to cars normally take a course of driving instruction. Would-be pilots need forty hours tuition before they qualify. Even the highly non-mechanical sailing boat demands expert tuition before a novice can handle it. But for two-wheelers it is often thought sufficient for the purchaser to be shown the controls and then to be left to his or her own devices.

In some ways this is a reflection of the inherent safety and simplicity of the powered two-wheeler. And especially with machines as soundly designed and constructed as the Hondas the safety factor is very high indeed. Self-tuition is certainly possible, but such is the crowded state of our roads today that it is no longer the best way of learning a new art. And an art it is, the handling of a responsive motor-cycle.

In some parts of the country it is possible to enrol for a course in the R.A.C./A.-C.U. Learner Training Scheme. This consists of a series of a dozen riding lessons and a dozen lectures in a course which lasts twelve weeks. The instruction is given by members of A.-C.U. clubs—real enthusiasts, who love motor-cycling and are happy to pass on their "know-how" to the newcomer so that he, too, will become not only a proficient rider but an enthusiast also. For make no mistake about it, once you have become a real motor-cyclist no other form of transport will ever really supplant it in your affections. Even if you never intend to do more than ride to and from work you will find that every trip is an enjoyable outing.

All the initial instruction takes place on private ground. When the instructor feels that the pupil is able to handle his machine well enough the lessons continue on public roads, and at the end of the course it is possible to take a test for a proficiency certificate. Unfortunately this does not, as yet, absolve one from the necessity of taking the M.o.T. Driving Test, but possession of the certificate certainly gives one confidence in one's ability to pass the test first time.

As the training scheme is voluntary it does not operate in all parts of the country, but there is a fairly constant expansion and the Secretary of the R.A.C./A.-C.U. Training Scheme at 85 Pall Mall, London, S.W.1, is always prepared to put inquirers into touch with their nearest centre.

Where no training scheme centre is handy self-tuition must of necessity fill the gap. The danger here is that bad habits may unwittingly be formed, and they become increasingly difficult to break. The riding style which the learner evolves at this stage is likely to remain with him, substantially unaltered, throughout his riding life. It determines not only whether he becomes a safe and skilled rider, but also whether he is one who gives his machine an easy life or whether he becomes an habitual "wrecker." Much wear and tear on the bike—and a considerable amount of time- and money-consuming mechanical tinkering—can be saved by riding habits alone, providing they are good ones!

Start your studies in an armchair, learning all about your machine and its controls. Having memorized the location of each of them (including the various switches controlling the main and dipped beams, the winkers, the horn, and so forth) close your eyes and pretend you are on the machine. Then give yourself a series of quick tests, for example: "Apply the front brake," "Signal a right turn," "Dip the lights," and so on. Make the appropriate hand movement at the same time, so that you eventually operate the controls almost as quickly as you can think.

When you have memorized all the controls and have finished your course of armchair instruction go outside to the machine, sit on it, and go through the entire exercise again. But in each case *do it with your eyes shut*. The reason is simple. On the road you will not have time to look down to see where your controls are. The place for your eyes is on the road ahead. Your hands and feet must be able to find the appropriate controls unaided.

The next step is to learn how to start the engine and how to control it once it is going. Switch on the petrol and, for the first start of the day, use the choke until the engine starts. If the air temperature is low you may also need to close or partly close it for warm starts too. Keep the Honda on its stand and get astride the machine. Switch on the ignition, check that the gears are not engaged, and open the twistgrip throttle about an eighth of an inch. Kickstart by placing the ball of the foot on the starter and thrusting downwards in an arc so that the starter is taken progressively down to the end of its stroke. Don't just jab—the aim is to give a steady thrust. On all C.102 models you will need only to press the starter button and the engine should start almost at once. Let it warm for a minute or so and then open the choke fully. On the mopeds, of course, you must pedal.

Now sit on your machine and spend a few minutes gently opening and closing the throttle so that you become accustomed to the way the machine responds. Don't jerk it open—just revolve the grip smoothly a fraction of an inch at a time, noting how the engine speeds up as you do so. Don't allow it to race. As there is no load on it, the motor will speed up very quickly indeed, and a sharp burst of throttle could cause it to over-rev and damage the valve gear. All you need to do is to accustom yourself to the feel of the control and learn how to operate it smoothly to avoid jerky progress when you start to ride.

You are now ready to make your first excursion. If you have access to a private drive—or, in the country, if a farmer will let you into a fallow field—then by all means make use of it. If not, choose a quiet road where other traffic will neither endanger you nor be endangered by you. And, if it is at all possible, keep well away from houses so that residents are not annoyed by your short trips up and down the street. It is well worth while pushing the machine a mile or so to a suitable site if you are not sure enough of yourself to ride it there, rather than take risks at this stage.

One of your main difficulties will be mastering the delicate operation of the clutch and its co-ordination with the throttle—unless, of course, you have a machine fitted with an automatic clutch. But with all other models practise "take-offs" first of all. This can be done by standing astride the machine, both feet firmly on the ground, and starting the engine. Then pull out the clutch lever, engage first gear, and slowly release the lever again by unfolding the fingers of your left hand from the knuckles, not from the finger joints, for the first inch or so of lever travel. This is done fairly quickly, but as you feel the drive begin to take up and the Honda move forward check the action immediately. This condition, with the clutch still not fully engaged and therefore still slipping, is necessary during the initial moving-off period—otherwise the engine would simply stall as the full load was suddenly applied.

Withdraw the clutch again, select neutral, release the clutch, and wait for a few seconds before repeating the experiment. This ensures that you will not overwork the clutch during this essential practising period. Then go through the procedure again until you are certain that you can recognize the instant when the drive begins to take up, and can check your clutch hand movement accurately at that point. Once this initial movement has been mastered you should be able to make reasonably smooth get-aways each time.

With the automatic clutch models the problem is easier—you have merely to be able to recognize when the drive is taking itself up and get your feet onto the footrests as the machine moves away. Try a few standing starts by having the engine ticking over, engaging first or second gear, and then gently opening the throttle. As the engine speeds up the clutch will begin to engage and the scooterette will move forward. Get your feet up at once if you plan to ride on, or, if you are just practising, close the throttle and apply the front brake. There are no snags, but it is important to remember that the automatic clutch is coupled in three different ways and that one of these is to the gear-change pedal. Holding the pedal down with the foot will thus keep the clutch disengaged and releasing it suddenly can cause a most disconcerting "bunny-jump." *Always* release the pedal fully, immediately you have selected a gear.

The C.100, C.102 and C.50 also differ from the rest of the range in having only three-speed gearboxes and in the use of a positive "neutral" located between first and second gears. The theory is logical enough. Top and second gears give an adequate spread for all normal riding, including moving off. Use the machine as a two-speeder, and you have neutral readily available and easy to find. The neutral indicator light helps to ensure this too. In the unusual event of an extra-heavy load being carried, or a really steep hill needing to be climbed, you still have an "emergency low" gear beyond neutral. In practice the idea works well, though for optimum acceleration I have found that it is best to use the Honda as a normal three-speeder and make the necessary double change to go straight from first gear to second. A learner might be better advised to stick to the "two speeds plus" philosophy.

When you are certain that you can get your machine on the move without jerks and without screaming the engine at unnecessarily high revs you are ready for the next stage—your first excursion under power. Choose a quiet stretch of road, if possible, and get under way in first gear. Don't bother to change up—just concentrate on controlling your speed by use of the throttle. With the automatic clutch models, though, don't travel *too* slowly or you will find that the clutch will disengage.

Practise speed control with the twist grip until you can accelerate and decelerate smoothly. When you come to the end of your stretch of road either try your hand at a controlled turn under power or, if you are not yet sufficiently confident, get into neutral and wheel the machine round.

The next step will be to learn how to make your turn at the end of each run (don't forget to make your signals). Then you will be able to start using the gears as well. First of all try changing from first into second and back throughout the run. Then bring the other gears into use as well. Learn one thing at a time until you can start the engine, get away, engage top gear, decelerate, and turn smoothly.

With the motor-cycles, you will have to acquire the necessary co-ordination between hands and feet which alone will enable you to operate the gears and control the speed to a nicety. To change from a low gear to a high gear, for example, you will have to make several movements simultaneously—or virtually so. First, the throttle is snapped shut and the clutch is drawn out, the two hands—right and left—working as one. As the clutch is disengaged the left foot operates the gear control, and as soon as the gear has been selected the clutch is released and the throttle re-opened. The object is to make the change as quickly as you can.

To change down, the easiest method is to withdraw the clutch and, almost at once, operate the gear pedal. At the same time the throttle is given a quick blip to raise the engine speed and so ease engagement. Purists prefer to make a clean change by leaving the throttle opening constant and letting the engine speed rise as the change is made, judging the whole procedure to a nicety so that there is no noticeable change in engine note throughout. That is a skill worth acquiring as you gain in experience, but during this early stage it is perhaps better to use blipped changes.

Both when moving off and when coming to rest the place for your feet is on the footrests. It is a bad mistake to leave the feet to "trail" for an instant longer than necessary, since you are not properly balanced on your machine till your feet are up.

On take-off the feet should be raised as soon as the clutch begins to "bite," and, when halting, one foot (and one foot only) should be grounded just a fraction of a second before the machine actually comes to rest. Trailing feet are the mark of a motor-cycling illiterate.

Many learners, especially where only pedal cycles have been used before, make the mistake of using only the rear brake and ignoring the front brake. This is incorrect, for the front brake is the more effective of the two and it is far less likely to provide a skid under bad conditions. Skilled riders use the front brake almost exclusively for checking speed when making a turn, and for stopping they give it a slight lead over the rear brake.

The reason is simple. When the rear brake is applied the effect on the machine is to transfer the weight forward. The loading on the rear wheel therefore decreases, and so does its adhesion. By contrast, this weight transfer adds to the load on the front wheel and therefore improves its adhesion. Therefore try to use both brakes in unison. If you *have* acquired a rear-brake-only complex, break yourself of it at this formative stage by using only the front brake for making all practice stops (though both should be employed for emergencies, of course). This will enable you to get the feel of the front brake—though it should be remembered that both brakes will tend to gain in power once they have lost their initial newness and start bedding-in properly.

To brake hard, the initial movement is to flick the throttle shut and follow-through this right hand movement by grasping the front brake lever and applying gentle hand pressure. At the same time the right foot begins to press on the rear brake pedal. Both pressures are steadily increased as the machine slows, and as it comes to a standstill the clutch, too, is withdrawn. The result is a quick straight-line stop.

Great care must be taken when the roads are slippery. Under these conditions I prefer to slow against the compression of the engine and use the brakes only sparingly to "kill" the last few miles an hour. Slowing in this way involves nothing more complicated than closing the throttle and, possibly, dropping down through the gears. To keep the speed in check if I am descending a steep hill with a treacherous surface I also give just the barest tickle on the front brake. This must be done gently—the object is merely to prevent the machine accelerating, not to stop it.

A two-wheeler is steered by being leaned to one side or the other so that the balanced forces holding it in equilibrium are upset by the arrival of a third force. Unopposed, this third force would simply cause the machine to topple over, but the very act of leaning it forces it into a turn and a fourth factor—centrifugal force—then comes into play. The result is that the machine stays banked at an angle and goes smoothly round the corner. If a turn was initiated with the machine upright, there would be nothing but the friction between the tyres and the road to prevent centrifugal force dragging it bodily across the road away from the centre on the turn.

For each combination of radius of turn and speed there is thus an ideal angle of bank, but luckily it is not necessary to go to the trouble of having rate-of-turn meters and inclinometers to ensure that all is well—the rider can *feel* whether or not the machine is turning smoothly and can increase or decrease the angle of bank to suit the case.

Most of the work of cornering has, in fact, to be done before the corner is reached. The object—simply stated, if not so simply achieved—is to arrive at the corner at the right speed, in the right gear, on the right point on the road.

As the Honda comes towards a corner its rider has to weigh up the pattern of the road and decide how to tackle the turn. All the braking and slowing has to be done with the machine upright and this, therefore, is completed on the approach. The first stage is to lose just the right amount of speed and to get into the right gear. What gear this is depends on the circumstances and on the machine.

The approach to the corner is made in the selected gear, with the throttle either fully or partly closed. If necessary, the brakes can also be applied. It is heeled over to take the bend, and as the rider reaches the point when he can see all that lies ahead the throttle is opened gently and the machine taken through under power. As the model is straightened up the power is increased and the exit is made under acceleration and at a higher speed than was used on the approach.

Make your golden rule for cornering the one which every racing man learns by heart: "In slow; out fast." Otherwise, you may have an unpleasant acquaintance with the second half of that saying: "In fast; out feet first." And on the road there is even less need than in racing to take any risks.

3 Basic principles

METICULOUS attention to design, as well as superb production engineering, has made the Honda motor-cycles the top-class models that they are. But to obtain the best from his machine the rider, too, has got to put something into the common pool. And to make his contribution he must know what makes it tick.

Only four-stroke engines are produced by Honda. This term "four-stroke" refers to the number of working strokes in one complete cycle of operation of the engine. In the four-stroke power unit, a working cycle consists of four distinct strokes—in other words the piston travels from its uppermost position to its lowest, or vice versa, four times. Each cycle, therefore, comprises two downward and two upward strokes.

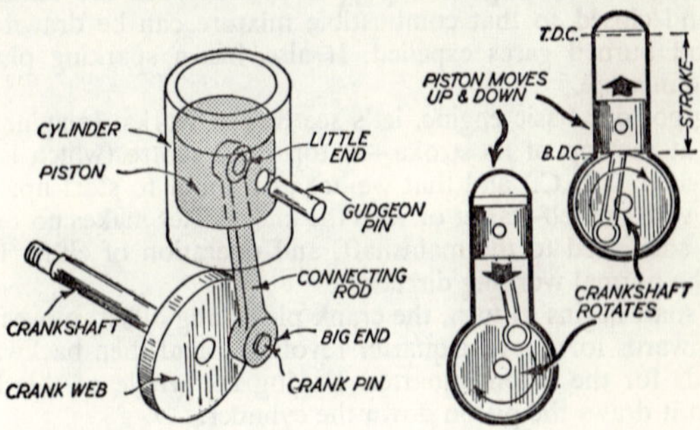

Fig. 6. The basic parts of a single cylinder internal combustion engine
(From *The Book of the B.S.A. Sunbeam and Triumph Tigress.*)

Before considering these in detail, however, let's get our bearings and assign names to the major engine components. The basis of the Honda engines—as of all reciprocating internal-combustion engines—is the crankshaft. This is basically a set of flywheels (two in the case of the singles; four in the twins) which are supported by main shafts projecting from the centre of each outer flywheel. These shafts are themselves carried on main bearings, mounted in an alloy container which is called the crank-case.

Joining the flywheels is a crank pin. This is not mounted concentrically, like the main shafts, but is set towards the outer edge of the wheels. So, if the wheels are revolved the main shafts merely go round and round. The movement of the crank pin, on the other hand, follows the circumference of a circle whose radius is the distance between the centre of the main shafts and the centre of the crank pin.

This "bottom half" assembly forms the rotating parts of the engine. But the power is produced, basically, by the reciprocating parts—those which move up and down. The two sections have a common member in the connecting rod, whose job is to transmit the reciprocating movement of the piston to the crank assembly for conversion to useful rotary movement.

The connecting rod is pivoted on the crank pin, friction here being reduced by a bearing which is known as the "big end." The reason is not far to seek—it is, in fact, the bigger end of the connecting rod. At the other end of the rod is another bearing—the "small end," through which passes the gudgeon pin. This is the pivot on which the piston is held.

Made of light alloy, with split cast iron rings to ensure a close fit, the piston spends its life moving up and down inside the cylinder, which is bolted to the mouth of the crankcase. Closing the other end of the cylinder is the cylinder head, which contains a hemispherical combustion chamber. In this are a pair of poppet valves which are automatically opened and closed so that combustible mixture can be drawn into the engine and burned gases expelled. It also has a sparking plug which ignites the mixture.

Having got our basic engine, let's see how it works. Imagine that the piston is at the top of its stroke—at top dead centre (which is usually abbreviated to T.D.C.) and that we are just about to start up. Whether it is done with the self-starter or with the kick starter makes no difference. Both are connected to the mainshaft, and operation of either turns the shaft in the normal working direction.

As the shaft begins to turn, the crank pin (being offset) moves forward and downwards for the first quarter revolution and then backwards and downwards for the second quarter. During the whole of this first half revolution it draws the piston down the cylinder.

All the time, the inlet valve has been open. As the movement of the piston has created a relatively low pressure inside the cylinder, the higher pressure of the outside atmosphere has forced a mixture of fuel and air through the inlet port and into the cylinder, so by the time the piston reaches the end of its travel—bottom dead centre, or B.D.C.—the cylinder is filled with inflammable gas. This stroke is therefore called the induction stroke.

It is not enough, though, for the mixture to be inflammable. To produce combustion which is fast enough for the purpose of driving an engine it must first be compressed. That is the purpose of the compression stroke, which follows. First, the inlet valve is closed so that the gas is trapped in the cylinder. Then—driven by the energy stored in the flywheels—the piston begins to rise up the cylinder. Crank pin movement during this second half revolution is upwards and backwards during the first quarter; upwards and forwards during the second, until at T.D.C. the crank pin has returned to its original position. By the time this has happened and the piston is at T.D.C., the mixture will have been compressed into a space perhaps only one-eighth or one-tenth as big as that which it originally occupied. Now, it will burn rapidly when ignited.

This is done at or about T.D.C. by an electrical spark which jumps over the points of the sparking plug. The resultant rapid burning causes the gases inside the cylinder to expand, exerting heavy pressures on all the surrounding surfaces. Of these, only one can move—the top of the piston. Consequently, the gases drive the piston rapidly down the cylinder and the connecting rod, now driving the flywheel instead of being driven by it, thrusts on the crank pin and so turns the main shafts under power. This is the power stroke.

As the piston approaches the end of this stroke the exhaust valve is opened and on the ensuing upward stroke the burned gases are driven out of the cylinder. This is consequently called the exhaust stroke.

This, then, is the basic cycle—induction, compression, power, exhaust.

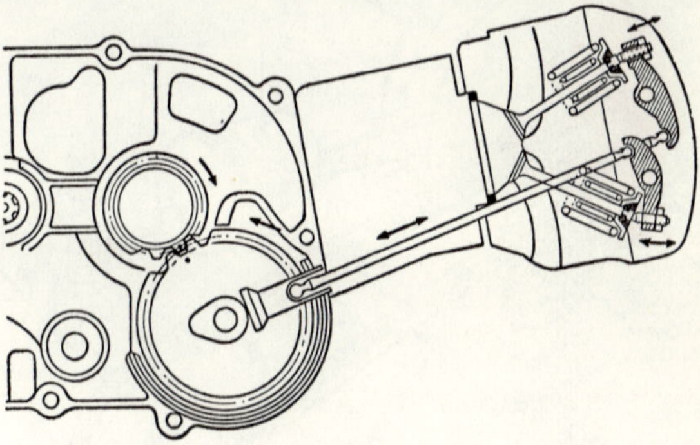

Fig. 7. The overhead valve layout. Honda C.100, C.102, C.110 and C.114 power units use overhead valves, with push-rod operation to open them. How it works is shown in this diagram

Towards the end of the exhaust stroke the inlet valve begins to open again, and a new cycle of operations begins. In the Honda engines, at top speed, the cycle can be repeated up to 5,000 times every minute in each cylinder.

In practice, the strokes are not quite as well-defined as this. Though the gas streams are light they—like every other form of matter—still possess inertia. They take a few milli-seconds to get on the move, or to change direction once they are moving. Consequently, a certain amount of overlap is allowed for at the end of the exhaust stroke and the beginning of the induction stroke, and at the end of the power stroke.

As the piston approaches B.D.C. on the power stroke, for example, the exhaust valve is opened. True, this sacrifices a modicum of power, although there is very little energy left in the gases by this time. This sacrifice is more than recouped by the fact that the gases start to leave the cylinder under their own pressure. This reduces the amount of work that the piston has to do to expel them as the exhaust stroke gets under way.

While the piston is still travelling upwards on the last part of the exhaust stroke the inlet valve is opened. This does not result in the stale gas being pumped into the inlet port—it is already moving too quickly into the relatively low-pressure area of the exhaust pipe for that—but it *does* give the inlet gas stream a chance to get on the move before the induction stroke proper begins.

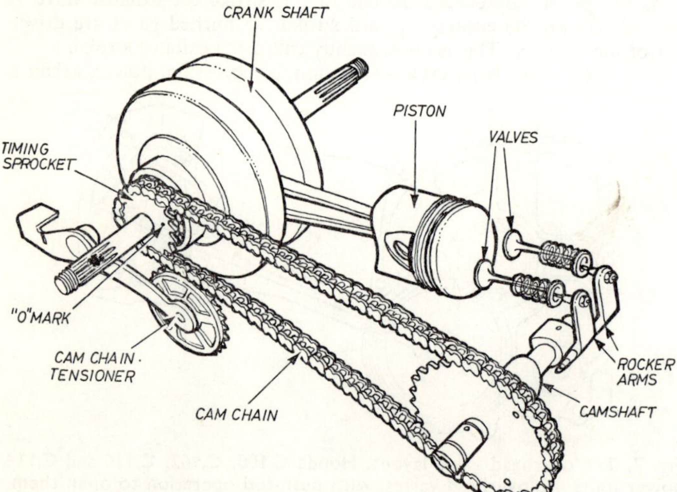

Fig. 8. The overhead camshaft layout. This basic design is employed on the P.50, PC.50 and C.50 models, the camshaft being mounted in the head and driven by a chain

All this, of course, argues very precise control over the valves. In the earlier 50 c.c. Honda engines the valves are operated through rockers and push-rods—an overhead valve arrangement. The later models use rockers on which the camshaft bears direct—an overhead camshaft arrangement.

In each case, the contours of the cams determine the behaviour of the valves. A cam can be regarded as a wheel which has been deliberately distorted so that instead of rolling smoothly it has an up and down action. Anybody who has ridden a bicycle with a wheel out of true knows what happens. As the machine moves along the front or rear end continually rises or falls as the unevenly radiused wheel revolves.

The action of a cam is similar, save that the camshaft is held rigidly in position and cannot move up or down. The movement is therefore transmitted to the surface which is in contact with the cam periphery—to a tappet block in the case of the o.h.v. motor, and direct to one end of a rocker in the o.h.c. engine. As the cam rotates these parts reciprocate. The tappet block transmits its movement to a pushrod which itself bears on one end of a rocker. In each case the rocker see-saws about a central pivot, so that the other end either depresses or releases the stem of its valve. These valves are, of course, spring-loaded, so that they are normally held tightly shut.

A further vital part of the four-stroke engine is the lubrication system. The lower part of the crankcase—the sump—contains a supply of lubricating oil, whose purpose is to reduce friction between the various components of the engine and to help absorb internal heat. Some of the oil is distributed, in the form of a mist, simply by the splash effect created by the revolving internal parts. But it is also essential that it is taken to the vital big-end bearings and to the valve gear. On the C.50 there is a mechanical oil pump. All the other models rely on splash lubrication. Driven off the lower run of the cam chain, the C.50's pump sucks in oil and drives it through internal passage to the more heavily-stressed areas. Its job done it is allowed to fall back into the sump, where it has a chance to cool again before being pumped through the system once more.

THE IGNITION SYSTEM

Often, even experienced riders have only the slightest knowledge of the working of the electrical system upon which the whole operation of the engine depends. As a result the electrics tend to be neglected until failure results, when the assumption is made that "electrickery" is an unreliable thing anyway!

However, there is no need to be a qualified electrical engineer to understand *how* the system works, even if the actual reasons behind it have to be taken for granted.

All electrical practice is founded upon circuits and upon the fact that an electric current will invariably take the shortest path to earth, flowing from the negative to the positive pole. It should be emphasized that "earth" does not necessarily mean the ground. So far as a Honda's electrical system is concerned "earth" is the mass of the machine itself.

A circuit is just what its name implies. In this, electricity is rather like a model railway train. If all the points are correctly set the train will go round and round. If they are not so set it will simply end up standing still on a siding. Or it may be routed on to a damaged line and so derail itself.

As with the train, so with electricity. Providing there is a circuit the current will flow; if the circuit is broken it will not. And a bad points setting—a short circuit—may direct it straight to earth.

Electricity is measured in volts and amperes. The volt is a measure of its electrical force; the ampere is basically a measure of the number of electrons per second passing a given point.

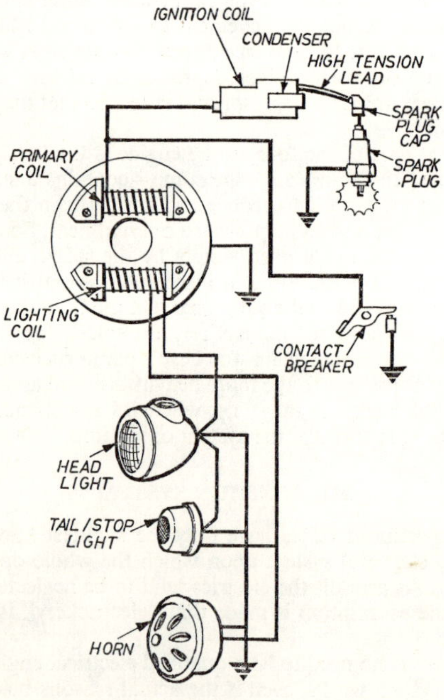

Fig. 9. A simple ignition and lighting system. This diagrammatic view shows the relation between the main parts of a direct ignition lighting system as used on the P.50, save that the stop lamp is not fitted to UK versions

So while voltage indicates the electrical pressure, amperage shows what quantity of current is flowing. The resistance to the current flow inherent in the wires and so forth which make up the physical side of the system is measured in ohms, one ohm being a resistance which calls for one volt to be applied so that one ampere may flow.

Finally, one must accept a single basic fact—that when a coil is placed within a magnetic field, and they move relative to each other, electricity is produced in the coil windings.

On most Hondas, the main electricity supply comes from an a.c. generator which is driven directly by the engine crankshaft. A battery —charged by the generator through the medium of a selenium rectifier— is provided to give additional current when the demand rises above the output of the generator.

The ignition system comprises the generator and battery, a form of mechanical switch known as the contact breaker (together with its associated advance/retard mechanism), a condenser, a coil, a sparking plug, and the associated leads and switchgear.

Primary current is supplied from the electrical system to the low-tension side of the ignition system. This comprises the contact breaker, the condenser, and the low-tension windings of the coil. These windings urround a high-tension winding.

Here is where the actual happenings inside the system have to be taken on trust. While current is flowing through the low-tension windings, there is none whatsoever in the high-tension system. But as soon as the low-tension current is broken—which is the job of the contact breaker—the result is to generate high tension in the secondary windings. In the case of the Honda this is in the region of between 15,000 and 20,000 volts. Seeking the shortest path to earth, it races down the heavily-insulated ignition lead, which is connected to the central electrode of the sparking plug.

Now the plug's central electrode would, normally speaking, be an electrical dead end, since it is not connected to earth or a circuit and is separated from the plug's side electrode by a gap of some 0·025 of an inch.

Electricity at low tension would, in fact, stop short at the gap. But in the high-tension system, with such pressure behind it, it cannot do so. Instead, it leaps across it in the form of a hot blue spark, and it is this spark which ignites the mixture.

As this happens some 5,000 times each minute in each cylinder it is obvious that pretty accurate timing is required. This is provided by the contact breaker, whose opening and closing is controlled by a cam on the engine's main shaft or on the camshaft.

Of the two contact breaker points one is carried on a plate, the other on a centre-pivoting arm. The far end of this arm has a heel which bears on the cam. Consequently, as the cam revolves the arm see-saws and the points open and close. To prevent the low-tension current from backfiring and bridging the points while the gap is small an electrical shock-absorber—the condenser—is added to the low-tension circuit. This absorbs unwanted pulses of voltage.

The C.100 scooterette has the simple flywheel magneto type of ignition in which the coils are contained within the flywheel itself, on a back plate which also bears the contact breaker and condenser. Here, the electricity is generated by the effect of the rotation of magnets fixed into the periphery of the flywheel and no provision is made for altering the timing which, in this case, is set to 35° B.T.D.C. The C.102 version, on the other hand, has a.c. type electrical equipment and automatic variation of the spark between 5° and 35° B.T.D.C.

THE CARBURETTOR

So far one vital part of the power unit has not been considered. We noted in passing only that when air is induced into the cylinder it is mixed with petrol to form a combustible mixture. And that was a drastic understatement of the work done by a simple but precisely-engineered instrument known as the carburettor.

In principle this may seem to be little more than a glorified scent spray with a fancy name. But it has to carry out one of the most exacting of all tasks—metering a minute quantity of petrol and mixing it thoroughly with air in just the right proportions for efficient burning.

At first this may not seem over-exacting, since the ideal ratio is roughly 1 part of petrol to 14 parts of air by weight. The carburettor, however, does not operate by weight but by volume. On this basis, each 50 c.c. of combustible mixture needs to contain only 0·005 c.c. of petrol. The remaining 49·995 c.c. is all air, so the proportion actually metered by the carburettor is nearly 1 part of petrol to 10,000 parts of air. Obviously, despite its simple design, a carburettor is a precision instrument and needs to be treated as such.

The basic components of a carburettor are a petrol reservoir, called a float chamber; a venturi, or choke, through which air is drawn; jets, which meter the petrol; and a throttle, which controls the amount of mixture passing through the instrument.

First consider the basic method of operation. Petrol is fed to the float chamber, which is much like a pocket edition of the domestic cistern. It contains a float (a double float in the case of the motor-cycles) which rises as petrol is admitted through a valve. In doing so the float presses the valve's tapered needle upwards. As this needle is carefully contoured to fit in the valve seat it gradually shuts off the fuel supply, and when the preset level has been reached no more fuel can enter the chamber until the engine uses a proportion of the supply already there and the level falls, the float sinks with it, the needle valve opens, and more fuel flows in until the correct level is again attained.

Connecting the float chamber with the body of the carburettor is a drilled passageway through which the fuel can flow into a jet well. Obviously, the level in the well is controlled by the level in the float chamber. Immersed in the fuel in the well is a jet. This is screwed into the end of a needle jet tube, the other end of which opens into the venturi.

This jet looks suspiciously like a small bolt with a hole drilled through its centre. In fact that is just what it is—but the hole is so proportioned that it will pass just the right amount of fuel and no more.

When the induction stroke begins in the cylinder air is drawn through the carburettor venturi. This is so shaped that as this air passes through there is a fall in pressure in the section into which the jet tube projects. As a result, petrol from the jet well rises up the needle jet, passes into the air stream as a fine spray, and mixes with the air in a section of the venturi called the mixing chamber. From there, it passes down the inlet tract and into the cylinder.

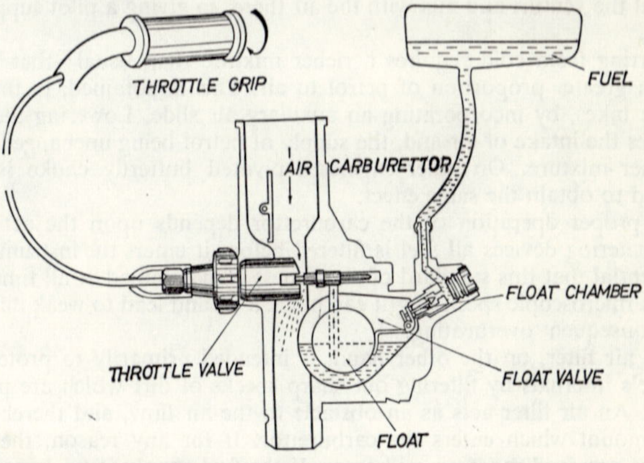

Fig. 10. How a carburettor works. Fuel from the tank is fed to the float chamber. It passes through a passageway ("jet") to mix with air entering the carburettor. The amount of fuel/air mixture passing into the engine is controlled by the throttle valve assembly, which is operated through the twist grip

Obviously, a carburettor which contained only these parts would work but would pass the same amount of fuel and air under all conditions. Consequently the engine would run at a fixed speed. Since the engine speed must be variable, so too, must the amount of fuel and air which enters through the carburettor. This is done by adding to the basic layout just described a throttle slide and a needle. The needle is clipped to the throttle slide, and projects below it to enter the needle jet. The throttle slide is barrel-shaped, and has a moon-shaped cutaway in its leading edge to govern the amount of air which can be admitted at intermediate throttle openings. The combined effect of this cutaway and of the tapered needle is to restrict air and fuel flow proportionately. The slide/needle assembly is so arranged that it is raised when a throttle control is opened and lowered as the control is shut. The more the throttle is opened the higher the slide is lifted and the more air it permits to pass. Since the needle is attached to the slide this, too, is lifted in the jet tube. The tube is tapered internally, and raising the needle therefore increases the effective volume of the tube, enabling a greater amount of fuel to pass into the venturi.

For idling only a small quantity of air is admitted through the cutaway and under the rear edge of the slide, which is held partly open by an adjustable throttle stop. The amount involved is too small for effective metering by the needle jet, and this is therefore closed completely. Instead, a minute quantity of fuel is by-passed through internal drillings to emerge behind the venturi and mix with the air there, so giving a pilot supply for idling.

Starting from cold requires a richer mixture than usual—that is one with a greater proportion of petrol to air. This is obtained, in the case of the bikes, by incorporating an auxiliary air slide. Lowering the slide reduces the intake of air and, the supply of petrol being unchanged, gives a richer mixture. On other engines a pivoted butterfly choke is used instead to obtain the same effect.

As proper operation of the carburettor depends upon the action of fine metering devices all fuel is filtered before it enters the instrument. It is essential that this standard of cleanliness is maintained at all times, for even a microscopic speck of grit can block a jet and lead to weak mixtures and consequent overheating.

The air filter, on the other hand, is intended primarily to protect the engine's internals by filtering out sharp specks of dirt which are present in air. An air filter acts as an obstacle to the air flow, and thereby cuts the amount which enters the carburettor. If for any reason, then, the filter is removed more air will enter. If the fuel supply is unchanged the result, again, will be a weak mixture. Damage to the engine interior apart, this is one good reason why the motor should never be run without its cleaner, unless the jets, etc., are suitably modified to suit the new conditions.

THE CLUTCH

A Honda clutch consists of a series of plates. These are alternately plain and friction-lined. And alternate plates vary in the manner of their fixing inside the clutch body. Some of the plates have plain circular centre cutaways, but are splined on their outer periphery. The others have plain circumferences, but have splines projecting into the centre cutouts. The former type mate with slots cut in the drum-like clutch body; the latter mate up with splines on the clutch centre.

The clutch body is driven by the engine mainshaft on which it is mounted. The clutch centre carries the drive gear. It is the job of the plates to form a link between the two. Strong springs hold the lined and unlined plates together. When the clutch is home, the friction between the driving and driven plates is such that they revolve together. The drive therefore reaches the gearbox through the combined motion of the clutch body and centre, transmitted through the plates.

When the clutch is operated the pressure on the springs is relieved and there is insufficient friction between the plates to enable any movement to be transmitted. Instead the plates which are splined to the clutch centre merely remain stationary while the adjacent plates which are splined to the clutch body freewheel. Gradual release of the clutch lever brings the plates together progressively, and the drive is therefore taken up smoothly. This happens whenever the machine is moved away from a standstill. When changing gear freeing the clutch momentarily relieves the gears of loading and enables them to move easily.

On the C.50, C.100 and C.102 scooterettes the clutch is controlled centrifugally, and is also coupled so that it disengages momentarily as the foot-change pedal is operated.

THE CYCLE PARTS

When a motor-cycle is driven along a road it remains upright for exactly the same reason that a gyroscope refuses to topple over—its two revolving wheels act, in effect, as a pair of gyroscopes, and resist attempts to force them out of their course.

There are, however, other factors which enter into it. One is the design of the steering gear. This is so arranged that, although the fact is not immediately apparent, the front wheel is trailing, rather like the castor of an armchair. To some extent the steering characteristics depend upon the amount of trail specified by the designer, and upon certain other factors such as the rake of the steering head, the weight distribution of the machine, and the correlation of the centre of gravity with the roll centre. And, of course, the efficiency or otherwise of the suspension systems plays a considerable part. All the Honda machines dealt with in this book use a leading link front fork which gives the minimum of wheel-base variation.

At the rear all Hondas except the P.50 use a swinging-fork layout controlled by two spring units. Hydraulic damping is employed at both front and rear. This is designed to eliminate oscillation which would otherwise result from the action of the suspension springs. A hydraulic damper consists of an oil chamber formed in one of the two members of the telescopic unit and a disc valve carried on the other. On a bump the oil is forced past the valve and offers little or no resistance to movement. On the return stroke, however, the action of the valve is such that the oil has to pass back through restricted channels in the piston of which the the valve is part. The resulting drag slows down the rebound action of the forks.

BRAKES

Just as important as making a motor-cycle go is the ability to make it stop. On the Honda this is the job of a pair of internal-expanding brakes.

On the 50 c.c. models (except the P.50, which has an internal contracting brake at the rear) both front and rear brakes are of leading-and-trailing shoe internal-expanding type. In these, a pair of shoes, carrying friction

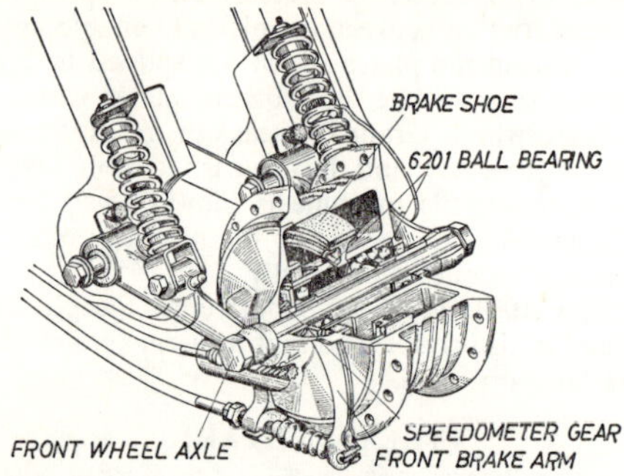

Fig. 11. A typical Honda front suspension/brake layout. This cut-away shows the P.50 design, which is virtually a miniature version of the layout used on the bigger machines

linings, is arranged concentrically inside the brake drum. One end of each shoe rests on a common pivot, and interposed between the other ends is a cam, operated through an arm by movement of the brake lever or pedal.

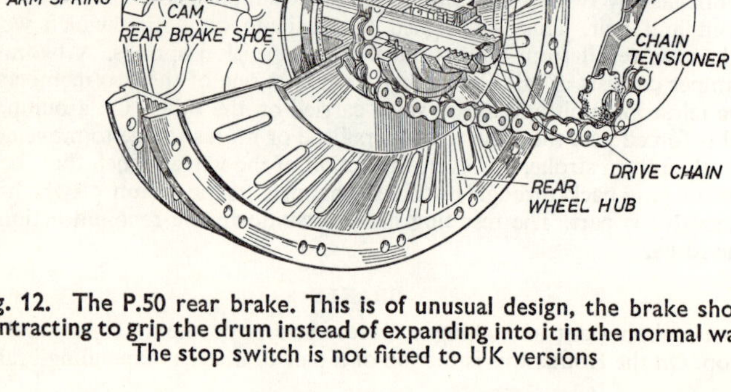

Fig. 12. The P.50 rear brake. This is of unusual design, the brake shoes contracting to grip the drum instead of expanding into it in the normal way. The stop switch is not fitted to UK versions

When the brake is applied the arm rotates the cam, which forces the shoes apart until their friction linings meet the drum. The resistance which they offer to rotation absorbs energy, and the machine stops. As the shoes are spring-loaded together, releasing the brake causes them to snap back into position.

4 Tools

Nobody could make a bigger mistake than to embark on serious maintenance work without an adequate set of tools—especially on a piece of engineering made with the precision of a Honda. True, Honda excel most other makes in the size and the quality of the tool kit supplied with the machine. Even so this is intended more for running repairs and simple routine maintenance than as a comprehensive kit to enable you to strip your motor-cycle, and for some jobs special tools are advised. This is not because the manufacturer thinks he can make a good thing out of selling them—in fact special tools are normally pretty uneconomic from his point of view—but because the machines are built to such fine engineering limits that this equipment is essential if parts are to be removed without damage.

To take a specific example, for the C.100 you need to have a metric cross-head screwdriver; 10 mm and 14 mm open-enders, rings or sockets, a 17 mm socket spanner, a flywheel extractor, and a clutch nut spanner.

Except on the PC.50, the Honda "singles" use nuts and bolts of 10 mm, 14 mm, 17 mm and 21 mm across the flats. These are the sizes you need in your kit. The PC.50 uses the new I.S.O. standard sizes instead.

When buying, purchase good tools. Cheap tools are a bad investment. Not only do they wear out quickly but they also have an infuriating habit of ruining nut and bolt heads as they do so. If you intend to become a *real* rider-mechanic get yourself a tool kit which will also play its part. A set of good metric chrome-vanadium open-enders is essential. They are sold in sets containing half a dozen spanners, and these will give you an ample range of jaw sizes. Complement these with a set of metric ring spanners—although these are not as handy as are open-enders when you have to get them into confined spaces, they have the outstanding advantage of being unable to slip off a nut.

A socket set can be regarded as a luxury—though if you ride a Honda, you will probably think, as I do, that such a magnificent machine deserves the best in the way of tools to go with it. And there is no doubt that a comprehensive socket set *is* the best, even though it cannot wholly replace either open-ended or ring spanners. Used with a variety of extension bars, however, sockets are both versatile and a hundred per cent safe. You can even obtain a universally-jointed extension which enable the sockets to be used round corners, and it is possible to obtain such desirable extras as ratchet or torque wrenches to go with them. For the average rider, however, a good stout set of box spanners will probably fill the bill almost as well. Made of steel tubing, box spanners can exert a considerable force if used with a sturdy tommy bar. It is essential, however, to buy only well-known brands, for cheap box spanners soon revert to being mere tubes!

Your tool kit already contains a choice of screwdrivers, but it is no bad plan to equip yourself with a pair of electrical screwdrivers for specialist electrical work—one of them short-bladed, the other with a long, partly insulated blade. For this, too, a pair of pliers incorporating wire cutters and strippers is essential. You will also need a set of feeler gauges, a collection of tin boxes with lids for the storage of small parts which are removed during overhauls, a valve spring compressor, and a grease gun. This last-named should be of the high-pressure type which can develop terrific pressures—over 8,000 p.s.i. have been recorded.

Make sure that you have proper surroundings in which to work. A machine as intricate as the Honda cannot be stripped in a road or a passageway. You need working space and plenty of light. An adjustable lamp, or one of the cheap clip-on lead lamps, will be helpful.

USING YOUR TOOLS

There is far more to the use of even the simplest of hand tools than merely placing them into position and tugging hard till the nut frees. Each particular type of spanner has its own characteristics and, generally speaking, for any one job you will find there is one type of spanner better suited for it than any other.

Unquestionably the great all-rounders of the tool kit are the open-enders which are slimly built and can, therefore, be slipped into confined spaces which no other spanner could reach. You will notice that their jaws are angled. This means that the open-ender will successfully loosen a nut even when you cannot obtain sufficient purchase to make more than a few degrees turn at a time. When the limit of movement has been reached it is possible simply to reverse the spanner and so obtain room in which to turn the nut a little more. In this way it is possible to undo a recalcitrant nut by easy stages.

It is, of course, essential that only the right size of spanner should be used. The open-ender is designed to apply its pressure along the flats of a nut or bolt and is consequently made with jaws of just the right width to grip them. If too large a spanner is used the jaws will press against the angles instead of on the flats, when one of two things can then happen. Either the spanner will gouge away the angles to leave a useless rounded head, or else the bolt head will prevail and spring open the jaws of the spanner. In the first case no spanner will in future be of the slightest use on that hexagon; in the second the spanner itself becomes just so much scrap.

Damage to the jaws can also be caused by applying excessive force when trying to free a bolt which refuses to budge. There is always a temptation to deal with the stubborn thing by slipping a piece of piping over the spanner to increase the leverage, or by locking the jaws of a pair of spanners together. The best advice is—don't. Sometimes these methods will succeed, but normally you will only apply excessive force and ruin the spanner. And there is always a danger of breaking the bolt completely.

Here is where socket, ring, or box spanners come into their own. Rings and sockets do not grip on the flats of the head but on the angles, and consequently force is applied all the way round. Boxes bear both on flats and angles. Obviously they have the advantage over spanners which can grip only on two surfaces. Consequently the same hand pressure applied through a ring or socket as was exerted through an open-ender will often free a nut which had previously resisted. Box spanners may, however, not prove quite so successful since the tommy bar may bend and unless a considerable downward force is applied the box may also tend to ride off the nut.

When using a spanner to tighten nuts and bolts it is important to remember that excessive force should not be employed. Spanners are designed to exert just sufficient force with hand pressure to tighten up most nuts and bolts correctly, and if excessive force is used the result may be to snap the bolt completely. At best the threads may be seriously strained. An exception is made only when definite instructions are given by the manufacturer—usually in the form of torque settings.

It is particularly important to bear this in mind on a machine such as the Honda in which a considerable amount of light alloy is used. The steel bolt is so much harder than the material into which it is threaded and over-enthusiasm with the spanner will merely rip the threads inside the hole and the bolt will come out.

Pliers, of course, should never be used as makeshift spanners since the jaws can never be parallel and the serrated pipe grip is perilously liable to slip. A rounded hexagon is the inevitable result if it does.

Neither should adjustable spanners be used, save in an absolute emergency. Here again it is impossible to set the jaws to the required accuracy, and rounded hexagons result.

Screwdrivers must have their blades properly ground so that, in side view, the blade is at first concave and then runs parallel as far as the tip. This enables it to be seated properly in the slot, and the pressure it applies is then equally distributed along the slot sides. A screwdriver with a wedge-shaped blade in side view does not seat properly and instead of an even pressure all the force is concentrated on the edges of the slot. These crumble and the screw is ruined. For cross-headed screws use only cross-headed screwdrivers. These come in two sizes and in two grades of hardness. Stick to those supplied in the Honda tool kit and you will be on the safe side.

After use, all tools should be wiped clean. They should be kept lightly oiled if they are used infrequently, and in any case wrap them in clean and dry rag (*not* plastics as these cause condensation and lead to rust). Before using them again wipe off the oil film or the tools may slip when any pressure is applied.

5 Be your own doctor

When a doctor wishes to diagnose a patient's illness he works methodically, listing the various symptoms to build up an overall picture of the complaint. Then, with the help of his knowledge of the way the body works, he can identify the illness and give the appropriate treatment for the disease.

Exactly the same procedure has to be followed when a motor-cycle refuses to work. Obviously there is a fault—some reason why it won't work—and before the fault can be cured it has got to be located and identified. The search has to be just as methodical in its way as is the doctor's.

Take the engine first. Assuming that certain basic requirements are met the engine *must* work. If it is not working it is proof that one or more of these requirements are not being met, and fault tracing is discovering which and why.

An engine must work if the correct charge of petrol/air mixture is being induced into the cylinder at the right time and is properly compressed and fired by a spark occurring at the right time, and the residue properly exhausted.

The first stage must be the obvious one of checking that there is, in fact, a supply of petrol reaching the carburettor. The first thing to do—surprisingly very often overlooked—is to remove the tank cap and make sure that there is some petrol. That done, and assuming that there is ample fuel, check that the petrol tap is switched on. If the supply is low, switch over to the reserve position (on those models which are equipped with reverse taps).

Taking the check one stage further, check whether the petrol tap or the fuel line is blocked, or the needle valve in the float chamber is jammed. To locate the seat of the trouble switch off the fuel, detach the fuel pipe at the float chamber end, and switch on again, when fuel should flow freely from the pipe. If it doesn't then obviously the blockage is in the pipe or the tap. Switch off once more, detach the pipe completely, then switch on again. If fuel flows through the tap then the pipe is the culprit—and that should be easy to clear. But if there is no fuel flow then your trouble is in the tap, and it will therefore need to be stripped for cleaning.

Should your first check on the fuel lines indicate that petrol is reaching the end of the pipe but not entering the carburettor it shows quite clearly that the fault lies in the float mechanism. The remedy is to remove the carburettor float bowl, again switch on the petrol, and watch the action of the float-controlled needle-valve. Frequently, working the valve with the fingers will free whatever obstruction had caused the blockage, and the flow of petrol which follows will then be sufficient to flush the value. In obstinate cases the only solution is to remove the valve and wash it in petrol. If the needle shows signs of damage replace the valve assembly with a new one.

It is possible for the fuel system to be at fault by supplying too much fuel, as well as by supplying too little. Overflooding, as this type of trouble is called, is depressingly easy to recognize—fuel pours from the overflow pipe whenever the tap is switched on and the engine, if it runs at all, constantly misfires and has a lumpy exhaust note.

Only the float assembly can be responsible for this particular form of trouble. A float may be punctured, in which case it sinks to the bottom of the chamber and allows the valve to remain almost fully open. But it is more likely that the answer is that dirt has entered the needle valve and is holding it off its seating. Even a tiny speck of hard matter is sufficient to prevent the needle from seating properly, and there is thus a constant trickle of fuel into the chamber. The effects of this milder form of over-flooding would be more noticeable at the lower engine speeds, when the excess fuel cannot be used up quickly enough. At high speeds the unit tends to take all the fuel it can get fairly happily, providing the critical petrol/air ratio of 12:1 (which gives maximum power) is not exceeded.

Akin to overflooding in its effects on the running of the engine is an incorrect petrol level in the float chamber. This is caused by the setting of the float having been deranged, so that the petrol level has to be higher before the needle will be pushed home. In all these cases stripping the system and examining and checking the components should uncover the fault.

Where initial inspection of the fuel system shows no immediately obvious fault the next stage of the fault tracing should be switched to the ignition system, and especially to the sparking plug. To a knowledgeable rider the plug can often tell a great deal about the conditions inside the engine. If the plug points and insulator are covered with soft black soot it is positive proof that the engine is running over-rich. On the other hand, if these parts have an ash-white appearance the inference is that the mixture is extremely weak, and one can begin to think in terms of blocked jets, or some other form of fuel starvation. Hard carbon on the plug shows that oil is being burned inside the cylinders—a sign of worn bores, or perhaps of faulty rings. But a plug which is coloured a deep coffee shade indicates that the mixture and combustion are normal.

Examination of the plug will thus give a good general guide to the engine's internal health and may offer some valuable clue which will enable you to trace the cause of the breakdown. But don't be misled into

2

allowing the plug elf to itsescape untested. To do this, leave the plug connected to the H.T. lead. Switch on the ignition, and then ground the plug body against a clean part of the cylinder so that there is an adequate earth. Angle the plug so that you can see the points gap clearly (first checking that the gap is as specified), and then operate the starter. A fat blue spark should jump across the gap. Repeat this test several times. You should get the same result on each occasion. Wear gloves, or you might get quite a nasty electric shock if the plug is not properly earthed.

If there is no spark you have a fault somewhere in the ignition system and your next step is to find out exactly where it is. First of all, eliminate the plug from your list of possible suspects by substituting for it a brand-new plug.

Whichever you use—new plug or proved plug—repeat the test just described. If there is a good spark to reward you then the obvious inference is that the old plug's insulation had broken down and the current was shorting straight to earth instead of jumping the points. Fitting a new plug in its place (after checking the gap) should thus completely cure the trouble. But if, on the other hand, the new plug fails as well then the fault lies somewhere between the plug terminal and the current source, and a much more exhaustive check will be needed to discover the cause of trouble.

Start with the plug cap. For a satisfactory test you will need a long pin or a nail or a piece of wire. Detach the plug cap—it screws into the lead—and into the hole bored by its screw threads thrust the pin. Then, with the ignition on, hold the lead by the insulated section so that the pin is about an eighth of an inch from the cylinder head and operate the starter. If a spark jumps from the pin to the head it proves that current is reaching the end of the H.T. lead but is not passing through the cap and into the central electrode of the plug. Snipping a quarter of an inch or so off the lead so that the cap screw bites into fresh cable may cure the trouble. If not, a new cap will be required.

Where no current is flowing in the H.T. lead turn your attention to the wiring. Check all contacts and all low-tension wires for security, and make sure that the H.T. lead's insulation is undamaged. Try the effect of wiring in a fresh length of H.T. lead, just in case the old lead has been fractured or earthed.

Using the wiring diagram as a guide, trace the wires from the ignition switch. The action of the switch itself is plainly visible from the operation or otherwise of the neutral lamp when the ignition is on and the gear is in the neutral position—if the lamp lights then the switch circuit is obviously in order. Lastly, examine the contact-breaker points and check the gap between them. If the points are dirty give them a clean and then try the ignition again. A roadside clean-up which is adequate for the purpose can be made by inserting a piece of stiff clean card between the points, closing them gently on it, and sliding it out against light resistance from the contacts. Repeated half a dozen times this will remove surface contamination. Check the gaps too, and rectify them if they are too wide or too narrow. Signs of severe burning of the points is evidence that the condenser has failed.

If all the checks still produce no reason why the ignition system should be inoperative the probability is that a major failure has occurred, and that the job will have to be referred to an agent who has special equipment for checking circuitry.

Complete engine failure for any reason other than ignition or fuel trouble is unlikely, save in the somewhat remote event of such a vital part as the camshaft snapping. Other troubles are far more likely to show themselves in the form of reduced performance or perhaps of erratic running.

One of the likelier causes of lack of pulling power is an incorrect tappet setting—and on Hondas the tappet clearances are very critical indeed. They *must* be properly set—and they *must* be checked frequently to ensure that the correct clearance is maintained.

There is a good reason for the clearance between the tappet and the valve stem. It enables the expansion of the engine when warm to be taken up. If the tappet was tight when the unit was cold the expansion of the parts would result in the valve being held away from its seating. The efficiency of the cylinder would thus be badly impaired for compression would be lowered, the working pressures reduced, and the valve seats would be exposed to the searing flame of the burning charge and would quickly deteriorate.

Where tight tappets are suspected it is possible to deduce where the fault lies by the way the engine behaves. If it is an inlet valve which is not being properly closed there will be a tendency for the engine to spit back through the carburettor, since some mixture will be driven back during the compression stroke and there will also be a leakage of gas during the power stroke. Where it is an exhaust valve which is not seating properly the mixture tends to be driven into the exhaust system and ignite there prematurely by virtue of its contact with the hot pipe. This gives rise to banging and rumbling in the exhaust. In both these cases the engine tends to run hot, and if the trouble is really severe it may cut out altogether at low speeds. The plug will of course reflect the trouble accurately enough on the inspection which should be done as a matter of course.

As it is far easier to rectify a tight tappet than to remove the head, recut the seats, and grind in new valves to replace burned or distorted ones, action should be taken without delay.

In cases of doubt it is better to err on the safe side and make a check on *both* the tappets. It is impossible to set them accurately with the engine hot, and most inadvisable to run with a tight tappet. As an emergency measure any tappet which is found to be tight when the piston is at top dead centre on the compression stroke should be slackened off until there is either no up and down play at all, or else just a barely perceptible amount of play. This is purely a "get you home" measure. When the engine is stone cold the tappets must be readjusted to the correct gap in the normal way.

Loss of compression can be caused in several other ways. If the unit has been run for any length of time suffering from chronic over-heating there is always a possibility that the cylinder head/cylinder block joint may have become distorted. In such cases, there will probably be a distinct hiss as gas escapes from the fractured joint on compression and exhaust strokes—or even an ominous cracking sound as hot gases bite their way out whenever the engine fires. Damage such as this calls for workshop treatment, since the head will need to be checked and refaced. Remedial measures should be taken as soon as possible because such a condition is one which gets worse the longer it is left. The reason is not far to seek. The leaking joint lowers the efficiency of the engine and allows air to dilute the mixture. This, in turn, leads to more overheating, which causes worse distortion.

Though the Honda engine is a sturdy and long-suffering unit it is utterly dependent upon regular oil changes—every 300 miles—and these should not be delayed or neglected. With use oils tend to lose much of their lubricating properties, and if a seizure should occur there is always a danger that the piston rings may be fractured. Besides losing compression the engine would then immediately start to burn oil, causing the exhaust system to smoke heavily. If ever this should happen, stop immediately—for the rings must almost certainly be broken and any further running will cause heavy scoring of the bores. The broken ends of piston rings make highly efficient cutting tools when propelled at internal-combustion engine speeds, so the cylinder could be virtually ruined.

Transmission faults are far more straightforward—either the drive is being transmitted properly or it isn't. But just occasionally what may seem to be a mild loss of power could be the beginnings of clutch slip, and where no apparent fault can be found in the engine it is always advisable to check on the adjustment and the condition of the clutch. Slip will be at its most noticeable if the throttle is opened sharply in the low gear on a stiff gradient. If the engine revs rise with no corresponding immediate improvement in traction it is a sure sign that the clutch is slipping, either through maladjustment or because the plates are wearing thin. No other cause is likely.

The process of elimination can also be employed successfully to isolate faults in the lighting system, bearing in mind the fact that if electricity is present and the circuit is complete then the electricity *must* flow. Where a circuit is dead it can be for only one of two reasons—either there is no electricity there, or else the circuit is wrong.

Faulty circuits are of two types—the open circuit and the short circuit. In the first case there is a simple break in the wiring. To use the railway analogy again, there is a rail missing. The train (our electricity) can therefore come up the line only as far as that point, and then halt. To get the train moving again it is necessary to repair the line. Electrically speaking, the whole of the wire from the current source to the break is still live but beyond the break it is dead.

The short circuit, on the other hand, can be regarded as a bit of faulty points-setting. Instead of the train being directed on to the correct line the signalman has diverted it on to a loop which leads back to the terminal. The train can continue to circulate on this loop line indefinitely, but it cannot take the right track until the points have been reset to allow it to do so.

As with the train, so with the electricity. Instead of a simple fracture of the wire we have a live conductor which has come into electrical contact with some uninsulated part of the machine. Since electricity invariably takes the shortest path to earth, our current has gone rushing off on this more attractive shorter route. In this case there is still a fully live circuit, but it is the wrong one! The useful part—the one that leads to one of the machine's components—is dead.

Before one can trace a circuit it is essential to have the appropriate wiring diagram and to be able to read it. At first sight this is a depressing prospect, for the average wiring diagram seems more like a small-scale plan of Hampton Court Maze. However, with practice it is possible to follow the circuits very accurately—providing the word circuit is always borne in mind. Part of the confusion arises from the fact that it is only the outward path of the current which appears on the diagram. The return is invariably through earth—which in this instance means the frame of the machine. Any components which are earthed to the frame are, therefore, automatically connected to one another electrically.

Where specific circuits have to be checked it is a good plan to obtain a sharp pencil and a piece of tracing paper and trace just that one circuit. With this data in your hand go to the machine and check through the various leads stage by stage. It helps to have a circuit tester—a small sharp pointer with a neon bulb in its handle and a clip by which it can be attached to the machine. Earth the clip, and press the pointer into the wire or on to contacts at intervals. If the bulb lights there is current flowing. If it doesn't, there is none. With the aid of such a tool circuit testing becomes relatively easy.

When the faulty section of the wiring has been located it should be closely examined so that the exact cause of the failure can be determined. A break inside the insulation can be pinpointed with the tester, or by simply tugging in opposite directions on the lead until you find a spot at which the casing merely stretches. This is where the lead has broken.

Where the suspect lead is a very long one, or is inaccessible, an alternative method of checking is to by-pass it with a long test lead equipped with small crocodile clips at each end. One is connected to each terminal and the circuit tested. If the hitherto inoperative component then works it shows that the fault is in the lead. In some cases it may be possible to draw a new lead into place by wiring it to the old one and then pulling it through from the far end, using the old lead as a guide.

To repair fractured leads so that no undue electrical stresses are set up the joint and the subsequent insulation should be made good as carefully as possible. If it is practicable to solder on proper snap connectors and encase them in an insulated sleeve this is much to be preferred. At the least the new joint should be well twisted together and bound over with two layers of insulating tape.

Terminals which have been undone must be properly retightened, and any new connectors crimped or soldered into place. Otherwise, high resistances and consequent early failure may ensue. Soldered joints are always preferred to plain ones, except in the case of the bared ends of wires clamped into a terminal post by means of a grub screw. Here binding the ends of the wire with solder is inadvisable since the grub screw may work loose in the resulting hollow and thus give erratic contact.

If these points are borne in mind there is no reason why the average owner should not be able to trace and rectify most minor faults on his machine, and provide at least a temporary cure.

6 Pinpointing troubles

Starting	Fault	Check	Possible Trouble
	Engine will not start or is hard to start	Is enough fuel reaching the carburettor?	
Sufficient fuel	Insufficient fuel		1. No fuel in tank 2. Fuel pipe obstructed 3. Fuel tap blocked 4. Tank cap vent blocked
		Remove the plug and earth it on the head. Operate the starter and check the spark	
Spark produced	No spark produced, or only a weak one		1. Faulty plug 2. Dirty plug 3. Dirty contact points 4. Points gap wrong 5. Plug gap wrong 6. Faulty condenser 7. Faulty H.T. lead 8. Faulty H.T. coil 9. Faulty ignition switch
		Operate kick starter to test compression	
Good compression	Poor compression or none at all		1. Starter not turning engine 2. Tight tappets 3. Valve not seating 4. Faulty head gasket 5. Burned valve head/seat 6. Worn or broken rings 7. Damaged head/barrel/piston
		Bump start the machine by pushing it in gear	
No sign of power	Engine fires but will not run		1. Choke opened too much 2. Air correction screw too far open 3. Air leak in inlet pipe joints
		Remove the plug once again and examine it	
Plug is dry	Plug is wet		1. Carburettor is flooding 2. Too much choke
		Start again with correct amount of choke	
Engine should run			

Running	Fault	Check	Possible Trouble
Lack of speed or pulling power		Lift and spin both wheels	
Both turn easily	Either does not turn easily		1. Brake is binding 2. Drive chain is too tight 3. Damaged wheel bearing 4. Hubs need greasing
		Check the tyres with a pressure gauge	
Tyre pressures are correct	Tyre pressures are wrong		1. Punctured tyre 2. Faulty valve 3. Neglected maintenance
		Engage first gear, apply the brakes and rev. the engine	
Engine stalls	Engine still keeps running		1. Clutch is slipping—adjustment wrong 2. Clutch plates are worn
		Slightly rev. the engine in neutral	
Engine speeds up	Engine does not speed up		1. Choke is closed 2. Air cleaner is clogged 3. Fuel line is clogged 4. Fuel cap vent blocked 5. Silencer is clogged 6. Main jet obstructed 7. Air leak in inlet pipe
		Test the machine on good roads. Then check the ignition timing and points gap	
Timing/gap correct	Timing/gap wrong		1. Defective adjustment
		Measure the tappet clearances	
Setting is right	Setting is wrong		1. Defective adjustment 2. Valve seat(s) in bad condition
		Operate the kick starter and test the compression	
Good compression	No compression		1. Valve seat(s) in bad condition 2. Rings and/or bore worn 3. Leaking gaskets 4. Valve timing incorrect
		Strip and examine the carburettor	
Carburettor is not dirty	Carburettor is dirty		1. Dirt is blocking jets
		Remove and examine the sparking plug	
Plug is clean and not discoloured	Plug is dirty or discoloured		1. Plug not properly cleaned 2. Plug is of the wrong grade
		Remove the oil filler cap and check the oil level in the sump	

Oil level is correct | Oil level is incorrect — 1. Oil level is too high, causing oil drag 2. Oil level is too low 3. Oil is dirty 4. Oil viscosity is wrong

- - - - Remove the domed head nut and check if oil is being pumped through when engine is running

Good flow of oil | No flow, or only a weak flow — 1. Oil pump inoperative 2. Oilway(s) blocked

- - - - Does the engine overheat?

Engine does not overheat | Engine overheats — 1. Excess carbon in the combustion chamber 2. Petrol of too low an octane rating 3. Clutch slips 4. Mixture is too weak

- - - - Test under high-speed acceleration

Engine does not "pink" or knock | Engine does pink or knock — 1. Petrol is of too low an octane rating 2. Piston, rings, and/or bore badly worn

Power should now be restored

IDLING FAULT	CHECK	POSSIBLE TROUBLE

Uneven running, mainly at low engine speeds

- - - - Verify the ignition timing and the tappet clearances

Timing/tappets are correct | Timing/tappets are incorrect — 1. Improper adjustment

- - - - Re-adjust the air correction screw on the carburettor

Original adjustment was correct | Adjustment found to be defective — 1. Mixture was too weak (tighten screw) 2. Mixture was too rich (loosen screw)

- - - - With engine running, spray petrol round the inlet pipe joints

Engine does not speed up | Engine speeds up — 1. Faulty O-ring 2. Distorted faces on pipe, head or carburettor 3. Faulty mounting of carburettor

- - - - Remove sparking plug, earth it on head, operate the starter and test the spark

Good spark produced | Poor or intermittent spark — 1. Plug faulty or dirty 2. Contact-breaker points faulty or dirty 3. Condensor is short-circuiting 4. Ignition coil is defective

Engine should now run evenly

Misfiring	Fault	Check	Possible Trouble

Uneven running at high engine speeds

- - - - Verify ignition timing and tappet clearances

Settings right

Settings wrong — 1. Adjustment incorrect

- - - - Disconnect the fuel pipe from the carburettor and test fuel flow

Normal flow

Weak flow —
1. Insufficient fuel in tank
2. Fuel pipe is blocked
3. Fuel tank cap vent blocked
4. Fuel tap is clogged

- - - - Remove the carburettor and examine the jet(s)

Jet(s) clear

Jet(s) blocked — 1. Blockage

- - - - Fit a new main jet and test engine

Satisfactory running is obtained

Running uneven —
1. Jet is too small
2. If trouble disappears with smaller jet either the air cleaner is fouled or the choke does not open fully

- - - - Replace the original main jet and check the valve timing

Timing correct

Timing is wrong — 1. Timing marks not matched

- - - - Strip the head and test the valve springs

Spring(s) in order

Spring(s) broken, or weakened — 1. Spring(s) defective

Running should now be even

Smoke	Fault	Check	Possible Cause

Excessive smoke from the exhaust

- - - - Run at continuous high engine revs.

No smoke

Coloured smoke produced —
1. Worn piston, rings, bore
2. Too much oil in sump
3. Rings fitted inverted
4. Damage to piston or bore
5. Metal fault in combustion chamber

- - - - Snap the throttle shut

No white smoke produced

White smoke is produced immediately the throttle is closed —

1. Inlet valve guide or stem is worn
2. Excessive clearance between head or guide

Run at continuous low engine revs.

White smoke is produced —

1. Exhaust-valve guide or stem is worn
2. Exhaust-valve guide is badly fitted

Engine should not now smoke

CLUTCH	FAULT	CHECK	POSSIBLE CAUSE
	Apparent power loss		
		Adjustment of clutch	
	Clutch slips		1. Weakened springs 2. Worn or distorted clutch plates 3. Worn or distorted friction linings
		Operate the clutch and engage gear at low speed	
	Engine stalls		1. Plates or friction discs distorted 2. Clutch springs of uneven strength

GEARS	FAULT	CHECK	POSSIBLE CAUSE
	Impossible to change gear		
		Clutch operation	
	Clutch operates perfectly		1. Gear-change drum dowel broken 2. Gear-change arm pawl damaged 3. Gear-change drum and fork movement defective 4. Selector fork broken 5. Second-gear dowel on output shaft broken 6. Second- and third-gear dowels on input shaft broken
	Gear pedal does not return		1. Broken return spring 2. Gearchange spindle binding on gearcase or cover
	Faulty selection or jumps out of gear		1. Dowel on second gear on output shaft worn 2. Dowels on second and third gears on input shaft worn 3. Selector fork bent or worn 4. Gearchange drum limit spring broken or weakened

Noises	Fault	Check	Possible Cause
	Tapping from region of cylinder head	Tappets	1. Excessive tappet clearance 2. Worn tappets
	Metallic knock or ringing from cylinder	Internals	1. Worn piston, rings, bore 2. Worn small-end bearing 3. "Pinking," caused by excess carbon in the head
	Rattle on o.h.c. machines	Cam chain	1. Tension wrongly set 2. Chain is stretched 3. Sprocket teeth worn
	Rattle from clutch side of engine	Clutch	1. Clutch drum slots worn, allowing plates to move 2. Clutch loose on its splines
	Rumble from engine	Crankshaft	1. Excessive end float 2. Worn or damaged main bearings

Steering	Fault	Check	Possible Cause
	Machine tends to skid or pull either way	Tyre pressures	1. Too hard 2. Too soft 3. Not adjusted for pillion passenger
		Operation of handlebars	1. Head bearings too tight 2. Steering race balls damaged 3. Stem is bent
		Vibration from front or rear wheels	1. Wheel bearings worn 2. Bent wheel rims 3. Spokes loose 4. Rear fork pivot and/or bushing loose or worn 5. Frame is bent 6. Faulty tyre 7. Rear chain incorrectly adjusted
	Machine tends to skid or to pull one side only		1. Unbalanced dampers 2. Wheels out of line 3. Front fork bent 4. Rear fork distorted 5. Front spindle bent 6. Loose head bearing

SPRINGING	FAULT	CHECK	POSSIBLE CAUSE
	Unsatisfactory operation of suspension	Tyre pressures	1. Too hard 2. Too soft 3. Not adjusted for pillion passenger
		Too hard	1. Front or rear damper not operating
		Too soft	1. Weakened spring 2. Load is too great
	Noisy operation		1. Friction between the fixed and moving parts of the damper casing 2. Friction between the casing and the spring 3. Rubber limit stop damaged 4. Hydraulic fluid level incorrect at front or rear

BRAKING	FAULT	CHECK	POSSIBLE CAUSE
	Brakes lack power	Adjustment at front and rear	1. Front brake cable binds 2. Brake rod is loose 3. Contact between shoe and drum defective 4. Water in the drums 5. Oil or grease on the linings
	No adjustment possible		1. Linings are worn 2. Operating cam worn 3. Shoe worn at point of contact with cam
	Noisy operation		1. Wear on linings 2. Dirt on the linings 3. Lining faces rough 4. Operating-arm bush worn

CHAIN	FAULT	CHECK	POSSIBLE CAUSE
	Chain needs frequent adjustment	Sprocket teeth	1. Wear on either or both sprockets 2. Teeth on either or both sprockets hooked 3. Wrong sprocket fitted
		Sprockets are serviceable	1. Adjustment neglected and chain has stretched 2. Periodic lubrication neglected

7 Methodical maintenance

JUST as, in everyday life, it is more important to keep fit than to be forever undergoing medical treatment and enduring major surgery, so with a Honda it is more important to keep it in good running order than to be continually stripping and rebuilding it.

On the face of it this may seem obvious, but a surprisingly large number of owners overlook the point and their engines tend to spend almost as much time in pieces as in action. Small defects neglected inevitably become major defects and these call for a stripdown to rectify the damage, and the stripdown in turn disturbs parts which have become bedded-in and settled. For a while the efficiency of the entire unit suffers.

Given the specified regular maintenance, any one of the Honda models can cover a quite surprising mileage before there is any real need to strip the power unit or even to have the head off for a "decoke." If, however, the routine maintenance is neglected the time which can elapse between overhauls is drastically shortened—and so is the life of the machine. Additionally the amount of work to be done and the amount of money to be spent will increase.

This is because maladjustments tend to have a cumulative effect. The Honda engine is tough—remarkably tough—but it is a piece of precision engineering, too, and it depends on this precision for the excellent performance which it gives. Run it fifty miles or so with a tight exhaust tappet and, though the power will fall off, little harm will be done. But in the absence of a routine check the exact nature of the fault may go undetected for hundreds of miles and all sorts of troubles can then result from this one minor example of neglect. The hot, newly-ignited mixture in the cylinder (and it is hot—exhaust valves work in temperatures up to 600°F) plays on the narrow sealing areas of the valve and seat like a blowtorch, scorching and pitting the metal. The stem of the valve may also be damaged. Since there is then leakage on compression and on firing the engine cannot develop its designed power and so the speed and acceleration power drop while fuel consumption rises. Eventually nothing but lifting the head and fitting a new valve will make good the damage. And that's a pretty stiff price to pay for the minute or so saved by not carrying out a weekly tappet check.

It is not only engines which can suffer in this way. The brakes are an even more glaring example. So good are the anchors on the Hondas that there is a considerable temptation to use them and forget them. They will, it is true, go on working for months with no apparent deterioration. But they *do* deteriorate—so gradually that the rider just doesn't notice that it's happening. He adjusts himself to the slowly altering feel and to the decreasing effectiveness and never for a moment suspects that they are not pulling him up as quickly as they could some months before. It is not till an emergency arises—one which could have been avoided easily if the brakes were in good order—that the real truth dawns. It can be a dangerous, if effective, lesson. And, again, it is one which can be avoided merely by testing and adjusting the brakes at regular intervals. Constant and methodical inspection is the best way of preventing any such troubles. Normally, checks and adjustments are recommended on the basis of elapsed mileage. This is an excellent scheme—*if* you keep a detailed log book. But all too often it is done on a basis of chance recollection—and memory is notoriously unreliable.

TASK SYSTEMS

Task systems are fundamentally different. They were devised for the Forces so that vehicles were kept under a constant mechanical check and so that no fault could go undiscovered for significant periods.

Systems suitable for the Hondas can be used on daily or weekly checks. Which is used depends entirely upon the work to which the machine is being put. If, on average, you are covering a couple of hundred miles in a week then the daily system is to be preferred. On the other hand, if your Honda is used mainly for weekend excursions giving a weekly mileage of less than 150 then tasks on a weekly basis can be substituted.

Taking the daily system first, here are Task Lists for all Honda machines. They are designed to cover all major components, yet to carry out these recommendations should never involve the expenditure of more than 10 minutes in a single day. And in most cases only a couple of minutes will be needed, for the idea is to *check*. Adjustments are carried out only if your examination shows that something is amiss.

Daily Task System

Sunday: check adjustment of front and rear brakes; check freedom of action of brake controls; check security of nuts and bolts in braking system; check lubrication of brake cables and linkages.

Monday: check engine/gearbox oil level; check final drive chain tension.

Tuesday: check all exposed electrical wiring for signs of abrasion or fracture; check all electrical terminals for tightness; check operation of all lamps and switchgear; check battery.

Wednesday: examine tyre treads and remove trapped stones; check tyre pressures; check spokes for security; rock wheels and front fork to test for bearing play.

Thursday: on C.110 and C.114, check clutch cable adjustment; on C.50, C.100 and C.102 check that clutch frees properly.

Friday: check all nuts and bolts for security; check operation of throttle and choke controls; check fuel filter for clogging.
Saturday: check adjustment of tappets; check sparking plug gap and condition; check contact-breaker points gap and condition.

Alternative Weekly System

Week 1. Check engine/gearbox oil levels; check tappet adjustment; check sparking plug gap and condition; check contact-breaker points gap and condition.
Week 2. Check brakes for adjustment; check all controls for freedom of action and lubrication; check wheels for security of spokes and play in bearings; examine tyre treads and check tyre pressures; check rear chain adjustment.
Week 3. Examine all electrical leads for signs of abrasion or fracture; check all terminals for security; check battery; check operation of horn, lamps, and switchgear.
Week 4. On C.110 and C.114 check clutch cable for adjustment; on C.50, C.100 and C.102 ensure clutch plates are freeing; check fuel filters; check all nuts and bolts for security.

By employing this approach to routine maintenance of your Honda you will ensure that most of the major points are, by the daily system, checked at least once a week. Even allowing for a pretty substantial utilization of the machine this should mean that no fault could go undetected for more than about 300 miles—even if you failed to notice it while riding the machine. And most defects would be discovered almost before they had had time to develop.

The weekly system, obviously, is less foolproof. A month could elapse between a fault developing and its being discovered. Consequently I would not advise this approach unless the average monthly mileage covered is less than 500 miles.

Neither the daily nor the weekly scheme covers such obvious periodic items as oil changes and greasings. These still have to be carried out on an elapsed mileage basis—obviously, since the exhaustion of lubricant is a question of actual use—and once again it is only too easy to forget just when the last change was carried out. Here, a useful aid to memory is to stick to the machine a small piece of coloured tape, on which you write "Next oil change due at *x* miles." On the scooterettes, this can be stuck under the seat just by the filler cap, so that you see it each time you fill the tank. With the motor-cycles, affix it to the frame just inside the tool box or battery box so that you spot it whenever the cover is removed.

It is a great mistake to neglect these periodic oil changes, for the lubricant in the Honda has a really important job to do. These engines are revving at something like 9,000 r.p.m. If the thin film of lubricant which is interposed between the working surfaces should break down the amount of wear which can occur in a short time is incredible. Moreover, the oil has not just one job to do but several. It lubricates the engine. It also lubricates the gearbox. And it has to help keep the inside of the unit at working temperature, the heat which it absorbs being dissipated through the finned crankcase.

In use, the oil becomes dirty. Condensation inside the unit causes a certain admixture of water (for every gallon of petrol which is burned the engine produces a gallon of water, and not all of this is expelled through the exhaust) and sludge results. By the end of the specified oil-change period the oil is thus contaminated with dirt, water, and tiny metal particles and is no longer capable of doing its job properly. If it is left inside the engine the wear rate is stepped up and efficiency suffers. For this reason, it is vital to adhere to the oil-change periods which are specified by the manufacturers.

For much the same reasons only the recommended grades of oil should be used. These recommendations are the result of long and expensive tests by the factory and by the oil companies. I have seen some of these under way at oil research centres and, believe me, they are thorough. By the time they are over, the technicians know exactly what lubricant will give least friction and longest life in any given engine. So I, for one, gratefully accept their advice and never begrudge the few pence which it costs to keep the engine well supplied with clean fresh oil of the right type.

Oil changes are best carried out at the end of a run, when the oil is hot. The reason for this is pretty obvious. Heat makes oil more viscous—it will run far more easily than when cold. Consequently, removal of the sump drain plug leads to an immediate and strong flow of oil out of the engine, and this carries with it most of the impurities which have collected there during the miles since the previous oil change. One word of warning, though. Always remove the oil filler plug after taking out the drain plug, or there may be a tendency for internal suction to hold back some of the dirty oil.

In all cases Honda drain plugs are located centrally underneath the engine. On the single-cylinder machines the drain plug is the 16 mm headed plug found immediately under the curved section of the crankcase casting.

Frequently, these plugs are very tight and it is essential for a good ring spanner of the correct size to be available. If necessary, the end of the spanner may be tapped with a hammer to help loosen a tight plug.

The C.50 model has a centrifugal oil filter set in the clutch body which should be cleaned out every 6,000 miles.

8 Fettling your Honda

To keep the single-cylinder Honda engine running well only a minimum of normal maintenance is required. The tappets, the plug gap, and the contact-breaker gap should be checked as advised and adjustments made when necessary. Finally the oil should be changed at the recommended intervals.

Checking the Tappets: All Push-rod Models. With the Honda engines correct tappet adjustment is vital. Since the clearance is very fine indeed —only 0·002 to 0·004 of an inch—greater care has to be exercised in this job. The engine must be completely cold. If it is hot, or even only slightly warm, it will be impossible to make an accurate reading. As a general rule it is better to check the tappets first thing in the morning or when the engine has stood without being run for at least five or six hours.

Wipe the area around the screwed rocker box caps with clean rag, so that there is no possibility of dirt or grit entering the engine. Then remove both caps. Next bring the piston to T.D.C. on the compression stroke. This is found most easily by removing the flywheel cover on the left side of the crankcase and turning the engine over until the mark on the flywheel coincides with the mark scribed on the crankcase. The appropriate feeler gauge is then slipped between the tappets and the inlet and exhaust valves in turn to test the clearance. Though it should slide in easily there should be no perceptible up-and-down play possible. If the feeler is left in position you should not be able to either lift or depress the rocker.

Adjusting Tappets: Push-rod Models except PC/PF.50. Each tappet consists of a screwed rod locked in place on the rocker's adjuster end by a 10 mm lock nut. Before adjustment can be made this nut must be loosened. Don't overdo things though. It is sufficient to undo it only a few turns—just enough to release the adjuster so that the tappet can be moved up and down by revolving the adjuster in the rocker.

Where a tight tappet has been discovered, slacken off the adjuster until the feeler gauge slips easily into the gap. With a loose tappet, insert the gauge and tighten the adjuster. I find it best, incidentally, always to keep the spanner on the lock nut throughout this operation so that the setting can be locked down to enable interim checks to be made. Vary the setting of the adjuster until the gauge will just slip into the gap with no up-and-down play on the rocker. Then, keeping the adjuster absolutely still, retighten the lock nut to hold the new setting. That done, check the gap again—for it is remarkably easy to move the adjuster slightly when tightening the lock nut and this results in the setting being deranged. If this has happened loosen the lock nut a thread or so and readjust. When you are satisfied that both tappets are correctly set replace the rocker caps and the flywheel generator cover.

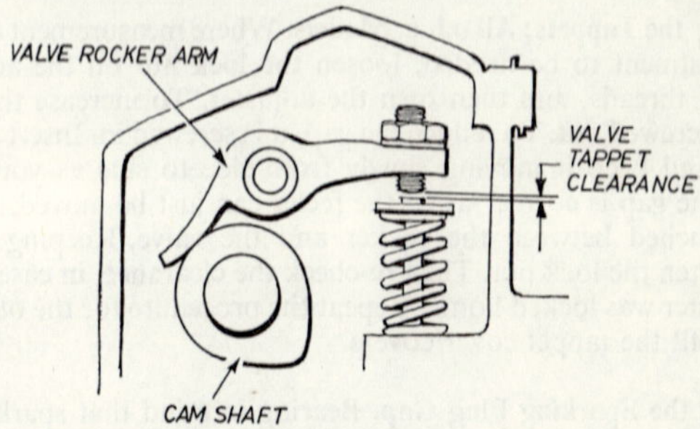

Fig. 13. Tappet adjustment. Maintaining the correct tappet setting is critical to obtaining full performance from your machine

Adjusting Tappets, PC.50 and PF.50 Push-rod Models. Though the method is similar, these engines have a simpler type of adjuster mechanism. Each rocker is held by two nuts, the outer one being a lock nut. Loosen this, and screw the inner (adjuster) nut clockwise to decrease the tappet clearance, anti-clockwise to increase it. Both valves are set to 0·002 in. cold.

Checking the Tappets: All o.h.c. Models. As with o.h.v. machines, it is essential that the tappets on the o.h.c. models should be kept within their specified working limits. They should be checked after the first 150 miles running, and again after 500 miles. Thereafter, measure the gaps every 1,000 miles. Both inlet and exhaust valve tappet gaps must be accurately set to 0·002-in. (0·05 mm) with the engine absolutely cold. The job is therefore best done first thing in the morning, after the machine has stood overnight.

The method of checking is to remove the dynamo cover and turn the flywheel until the "T" mark stamped on it aligns with the index mark scribed on the crankcase. You will probably find this easier to do if you first remove the sparking plug, to save turning the engine against compression.

With o.h.c. P.50 and PC.50s, detach the tappet adjusting cap on the cylinder head. With the C.50, take off the individual caps over the inlet and exhaust valves. Note whether the rockers are depresssing either valve. If so, turn the flywheel one complete revolution until the "T" mark and the index again coincide. This will place the engine at TDC on the compression stroke and free the valves.

Now measure the gap between the rocker and the valve stem, using a feeler gauge. If the correct gauge (i.e. the 0·02-in. feeler) fits the gap, try inserting the next-higher feeler as a double check. If this enters, the gap is too wide and must be reduced. On the other hand, if the correct feeler itself will not enter the gap is too small and it must be increased.

Adjusting the Tappets: All o.h.c. Models. Where measurement of the gap shows adjustment to be needed, loosen the lock nut on the adjuster by a couple of threads, and then turn the adjuster. To increase the gap the adjuster is screwed out. To reduce the gap, it is screwed in. Insert the 0·002-in. feeler, and keep it moving slowly from side to side as you turn the adjuster. The gap is correct when the feeler can just be moved, but is not actually pinched between the rocker and the valve. Keeping the screw steady, tighten the lock nut. Then re-check the clearance, in case it altered as the adjuster was locked home. Repeat the procedure for the other valve, and then refit the tappet cover/covers.

Checking the Sparking Plug Gap. Bearing in mind that sparking plugs work in a combustion chamber at temperatures many times greater than that of an oven, it is hard to believe that it is possible to maintain a gap between two metallic points with any degree of accuracy. In fact it is not only possible but vital. Though the plug will continue to work with a gap which is wildly incorrect it will not permit the engine to give of its best. Hence the importance of regular checks.

To check the plug, detach the pull-off cap carrying the H.T. lead and, using the box spanner provided in the tool kit, unscrew the plug from the head. On a new machine the plug is often a very tight fit, and under these circumstances it is permissible to tap the tommy bar sharply with a hammer to help the initial loosening. Using a feeler gauge measure the gap between the central and the side electrode of the plug. It should be 0·024/0·028 in. (0·6/0·7 mm.) If there is any significant variation from this, adjustment will be required.

Adjusting the Sparking Plug Gap. Gap adjustment is simple—it merely involves moving the side electrode nearer to or further from the centre electrode. Before doing so clean the plug. The best method is by sand blasting (a garage will carry this job out and set the gaps specified, for sixpence or so) but in the home workshop a stiff bristled brush usually has to stand in for the job. Do *not* use a wire brush. This can deposit metal traces on the nose insulator and allow current to leak away.

Brush the electrodes carefully, making sure that the bristles reach *underneath* the side electrode, and try to ease out any carbon deposits between the inner sides of the body and the nose of the insulator. That done, insert a feeler gauge into the plug gap and check the gap, for brushing away deposits may have altered it. If the gap is too wide leave the feeler in place and very gently tap with a spanner or the handle of a screwdriver on the top edge of the side electrode. This bends it slightly towards the centre electrode. By moving the feeler gauge in and out while you are doing this you will be able to feel exactly when the electrodes begin to bear against the feeler. The gap will then be correct. Where the gap is too small to allow the correct feeler to enter, the side electrode must be moved away from the centre electrode. It can be bent upwards by lifting it with a small screwdriver, or by use of one of the special gap-setting tools sold by accessories shops.

Never attempt to alter a plug gap by bending the centre electrode. This would inevitably fracture the ceramic insulator, and the plug will then have to be discarded. A fractured insulator would normally permit the current to short-circuit itself to earth and so prevent the engine running. On the other hand, the engine might continue to run while the broken section of insulator disintegrated inside. This would result in highly-abrasive particles falling into the engine with disastrous effects on the cylinder bore, the piston, and the rings.

Checking the Contact-breaker Gap: Models C.100, C.110 and C.114. First of all remove the flywheel cover completely. Then rotate the flywheel until the points—which can be seen through the inspection slots—are just opening. This position has to be found by inspection, and to aid vision here it is sometimes advisable to insert a thin piece of white paper (a cigarette paper is ideal) behind the points.

When this position has been found, the "F" mark on the flywheel should be aligned with the mark on the crankcase. The actual point gap is unimportant.

Model C.102. Remove the points cover, rotate the crankshaft till the points are fully open, and test the gap with a 0·014 in. feeler. As the contacts are fully exposed this is an altogether easier job than with other models in the range, and full visual inspection should be possible.

Adjusting the Contact-breaker Gap: Model C.102. When setting the gap between the contacts the short red line which is stamped on the face of the contact-breaker cam should coincide with the fibre heel on the moving arm of the breaker unit. Set the crankshaft to this position. Then release the two screws which hold the fixed contact plate. Do not undo these fully—merely loosen them sufficiently for the plate to move when the small eccentric adjusting screw is turned. Insert a 0·014 in. feeler gauge and alter the position of the eccentric with a thin-bladed screwdriver until the gauge is a sliding fit in the gap between the points. Then, holding the eccentric completely still, lock up the two fixing screws.

Subsequently reinsert the feeler gauge and check that the setting has not altered while you were tightening the screw. If it has, it will be necessary to carry out the procedure again to obtain exactly the right gap.

Checking the Timing: Models C.100, C.110 and C.114. Turn the flywheel *anti-clockwise* until the mark "F" on the flywheel coincides with the indented line on the flywheel casing. The points should now be just beginning to open—which means that a 1½ thou (0·0015 in.) feeler gauge can just be slipped between them. If it cannot it means that the points are still closed and the ignition timing is retarded—that is, the spark will be occurring too late. If the points are opening before the two marks coincide the ignition is over-advanced and the spark is too early.

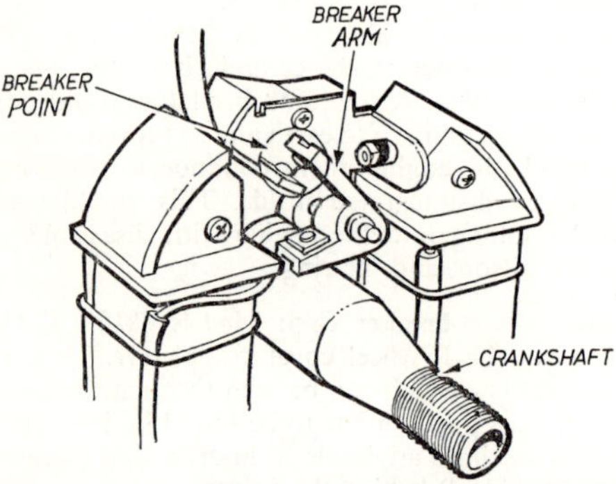

Fig. 14. Contact-breaker gap. The gap between the contact-breaker points varies the timing of the spark. Here, again, the working instructions must be followed carefully

In either case the points gap will need resetting. If ignition is retarded the gap must be increased slightly; if it is advanced then the gap must be reduced. After each adjustment, which is made with the points fully open as described above, turn the flywheel anti-clockwise to the two marks for a timing check. This is essential, for upon the correct timing depends much of the efficiency of the engine.

Contact-breaker Gap: Models P.50, PC.50, C.50. Although the gap between the contact-breaker points on these models should normally be between 0·012 and 0·016-in. (0·3 to 0·4 mm) the essential thing is that the *timing* should be correct—and this can only be varied by altering the contact-breaker gap. The normal contact-breaker gap check is, therefore, not made. Instead, check the timing itself as described here.

Checking the Timing: Models P.50, PC.50, C.50. Remove the left-hand crankcase cover and take out the sparking plug so that the engine is easily turned. Align the "F" mark on the flywheel with the index mark on the crankcase and examine the contact-breaker through one of the inspection holes in the flywheel. The points should just be breaking—in practice, should be open just enough to admit a 0·0015-in. feeler.

If you have a test lamp, you can check this even more accurately by connecting one of its leads to the black lead running from the engine

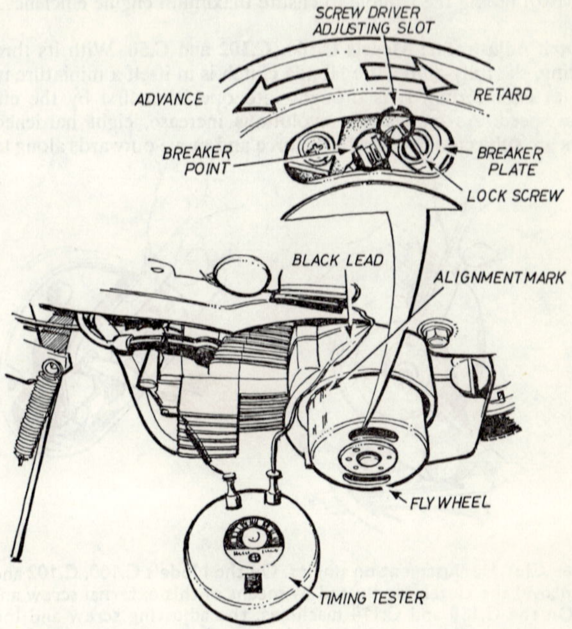

Fig. 15. Contact-breaker adjustment with test lamp. If possible, always use a test lamp when setting the ignition timing. This enables really accurate adjustments to be made

(break this lead at the connector) and taking the other to earth. Then rotate the flywheel. As the "F" mark on the flywheel comes up to the crankcase index mark your test lamp should show that the points have opened. If this happens *before* the two marks coincide the timing is over-advanced and the contact-breaker gap must be reduced. If the marks pass each other before the points open the unit is retarded and the gap must be increased. But remember that you *must* turn the flywheel in the normal direction of rotation, or you will get a false result.

To adjust, set the "F" mark and the index mark together, loosen the screw that locks the contact-breaker plate to the stator (just enough to allow the plate to be moved) and insert a screwdriver into the adjuster slot on the plate. Move the plate until the points are just breaking, tighten the clamping screw, and re-check. You may find it necessary to make one or two further fine adjustments to get the setting exactly right, but it is worth taking the trouble to ensure maximum engine efficiency.

Clutch Adjustment: Models C.100, C.102 and C.50 With its three-way coupling, the fully-automatic Honda clutch is in itself a miniature masterpiece of engineering. It is brought into operation first by the effect of engine speed. As crankshaft revolutions increase, eight hardened steel rollers are subjected to centrifugal force and move outwards along tapered

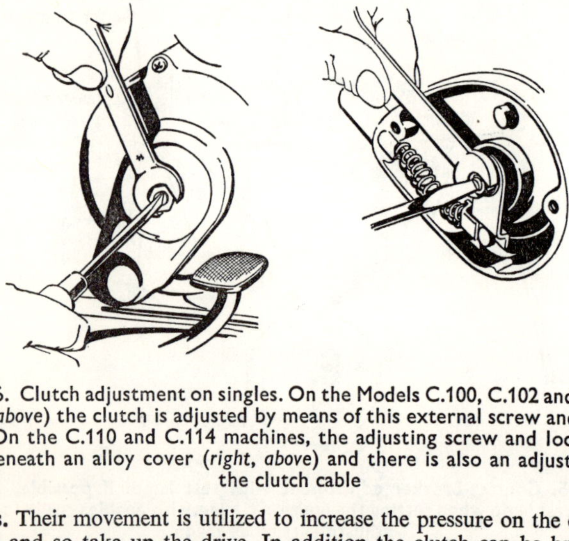

Fig. 16. Clutch adjustment on singles. On the Models C.100, C.102 and C.50 (*left, above*) the clutch is adjusted by means of this external screw and lock nut. On the C.110 and C.114 machines, the adjusting screw and lock nut are beneath an alloy cover (*right, above*) and there is also an adjuster on the clutch cable

tracks. Their movement is utilized to increase the pressure on the clutch plates and so take up the drive. In addition the clutch can be brought into operation by the action of the rear wheel over-running the engine when a gear is engaged—as, for example, in push starting. This same mechanism makes kick starting possible. It comprises a quick-acting three-start thread incorporated in an extension of the drive gear. Lastly, there is a direct connexion between the gear-change spindle and the clutch-withdrawal mechanism, so that the clutch is operated whenever the gear-change pedal is used.

Adjustment of the clutch is infrequently required. Unless there are unmistakable signs of clutch slip or drag, therefore, this component should be left untouched. However, when adjustment becomes necessary it is extremely simple to carry out. In the centre of the clutch housing on the right side of the engine will be found a screw and lock nut. Slacken the nut and turn the screw clockwise to move the clutch-operating cam plate away from the release mechanism. Full spring and roller pressures then operate on the clutch. Now turn the screw anti-clockwise until pressure can be felt on it. Stop turning immediately, and slacken the screw by a quarter of a turn in the clockwise direction. Still holding the screw, retighten the lock nut.

Models C.110 and C.114. These small Honda motor-cycles have a manually-operated clutch mechanism whose adjustment is extremely sensitive.

To reach the adjuster on the clutch-operating arm the pear-shaped cover on the right side of the crankcase must first be removed. This is held by two recess-headed screws. Now check that there is roughly a quarter of an inch of free play in the clutch control cable, and if not screw the cable adjuster in or out until this is obtained. You can then proceed with adjustment of the arm.

First slacken off the 10 mm lock nut on the centre adjuster screw and then rotate the screw anti-clockwise with a screwdriver by about an eighth or quarter of a turn. This ensures that there is no contact between the screw and the push rod which operates the clutch. Next rotate the screw clockwise until pressure is felt. Immediately, turn it back again in an anti-clockwise direction for just one eighth of a turn. Hold the screw steady with the screwdriver and tighten the lock nut to hold the new setting. Now make a fresh check on the control cable's free play. There must be just an eighth of an inch—no more and certainly no less. Use the cable adjuster on the crankcase to obtain this play. Finally, check that the clutch is freeing properly by kicking the engine over with the clutch disengaged. That done, release the clutch and kick the engine again to make sure that there is no slip.

Honda stress that when adjusting clutches on these models it is vital that the adjusting screw should *never* be screwed inwards against the pressure of the clutch springs. If this is done the plate which locates the three hardened balls will drop out of place. Should this happen, the entire clutch cover has to come off before they can be refitted.

Decompressor Cable Adjustment: Models P.50 and PC.50. The purpose of the decompressor control is to hold the exhaust valve slightly off its seat and so enable the machine to be pedal started with ease. Obviously, were the decompressor to be even partially in operation when the unit was running it would lead to a considerable loss of power.

To prevent this, the control lever on the handlebars must have between 0·04 and 0·16-in. free play (from 1 to 4 mm) before it begins to operate the valve.

This adjustment is made on the cable, which has a screwed adjuster fitted to the cylinder head. To adjust, loosen the lock nut and turn the adjuster—outwards to decrease play, inwards to increase it. When the control has sufficient free movement, retighten the lock nut.

On push-rod versions, allow ¼ in. play. Here, the adjustment is made at the solderless nipple on the actuating lever.

Carburettor Fuel Tap: Models C.100 and C.102. Both scooterettes are fitted with Keihin DP.13 HOV. carburettors of identical design—down-draught units in which the float chamber is held to the mixing chamber by a pair of countersunk screws, with the fuel tap incorporated into the float chamber top.

Very little trouble is ever encountered with these carburettors, but some cases of fuel starvation due to a faulty tap washer have been reported. If this is suspected, first obtain the necessary replacement washer (Part No. 16115-001-004) and then drain the fuel tank. This can be done by syphoning out the fuel with a length of plastic hose, or simply by removing the fuel line at the carburettor end and allowing the petrol to drain into a can.

Remove the front shielding. This is held by two bolts and a clamping nut on each side of the machine, and also by the air filter cover. This must be taken off before pulling the shield clear. To dismantle the tap remove the two screws which secure the tap cover and lift the cover, the spring washer, and the tap lever. Under the lever is the packing washer. Take out the old washer and discard it. Check that the fuel passages are clear and, if so, fit the new washer and reassemble the tap.

Carburettor Throttle Needle Adjustment: All Models. Generally speaking, little in the way of adjustment is either necessary or desirable on Honda machines, and it is possible to do more harm than good by inexpert tuning. For this reason, it is essential to note carefully by how much any adjustment is altered from the original factory setting. Then, if an experiment is carried out, it is always possible to restore the correct setting.

Misfiring at intermediate speeds is an occasional complaint, and quite often it proves to be due to an over-rich mixture. The first course is to check this. The composition of the mixture in the intermediate range of engine speeds (from 4,500 to 8,500 r.p.m. approx.) is controlled by a combination of throttle slide cutaway and the needle jet setting. Though the cutaway cannot be adjusted, the needle can.

To reach the needle screw off the top of the mixing chamber—the part where the control cable enters the instrument—and carefully withdraw the chamber top complete with return spring, slide, and needle. Detach the slide and needle by holding the chamber top and the cable, and pressing the slide upwards against the resistance of the spring. As the nipple on the end of the cable clears its seating in the slide, edge it to the outside and gently slide it through its slot. The needle/slide assembly is now clear.

A W-shaped needle clip plate holds the needle and clip in the slide. Ease this plate out by pressing the needle upwards *with your fingers*. Do not try pressing the needle on any other surface or there will be a risk of distortion. When the needle has been removed note in which slot the clip has been placed. Standard settings are in either the third or fourth notch from the top. Ease the clip off the needle and replace it *one notch*

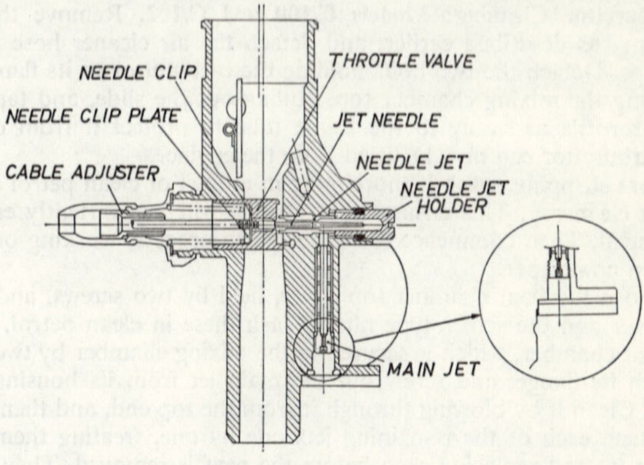

Fig. 17. The layout of a typical Honda carburettor. This is the type used on the P.50. The float chamber is not shown

nearer the top. This has the effect of seating the needle lower in the slide, and consequently weakens the mixture slightly in the intermediate ranges.

Replace the clip and needle in the slide, and spring in the clip plate. Offer up the slide assembly to the cable. You will find it possible to compress the spring and work the cable through its slot, finally easing the niople back into the seating. Wipe the slide clean with non-fluffy rag, and reinsert it in the mixing chamber. Remember that it is located by a peg, which engages in a slot in the slide and prevents it from turning. If the slide will not enter easily *don't* force it—it means that the peg is not engaging. Twisting the slide gently from side to side will allow the peg to take up in its slot. The whole assembly is then pushed home and the chamber top screwed down.

Where this procedure is unsuccessful, try the effect of renewing the air filter element and fitting the later-type "fat" air hose (C.100/C.102 models only).

Carburettor Tickover Adjustments: Models C.100 and C.102. Tickover is regulated by two screws on the carburettor body. They are the throttle stop screw, which alters the position of the throttle slide relative to the mixing chamber, and the air adjusting screw. The throttle stop is mounted centrally on the side of the body and the air screw higher and to the right.

The correct setting of the air screw is 1–1¼ turns out. With the screw so set the tickover speed is regulated simply by screwing the throttle stop screw in or out until a reliable tickover is obtained. It is best to do this with the engine hot and after checking that no air leaks exist at the flange joint.

Carburettor Cleaning: Models C.100 and C.102. Remove the front shielding, as described earlier, and detach the air cleaner hose and the fuel pipe. Detach the two bolts holding the carburettor to its flange after removing the mixing chamber top, pulling out the slide, and taping the entire throttle assembly to the frame tube to protect it from damage. The carburettor can now be lifted from the engine.

Before stripping, wash it thoroughly in a bowl of clean petrol and dry it with clean rag. This ensures that no dirt will inadvertently enter the instrument. Then commence the dismantling process, working on sheets of clean newspaper.

Remove the float chamber top cover, held by two screws, and lift off the cover and the screen-type filter. Wash these in clean petrol. Detach the float chamber, which is secured to the mixing chamber by two screws beneath its flange, and screw out the main jet from its housing in the flange. Clean it by blowing through it from the top end, and then replace it. Detach each of the remaining jets one by one, treating them in the same way, and replacing each before the next is removed. They are the slow-running jet, mounted on the side of the carburettor just off-centre of the horizontal mixing chamber, and the needle jet which is located under a plug concentric with the bottom of the mixing chamber. When rebuilding the carburettor, use new gaskets and washers throughout.

Carburettor Adjustment: Models C.110 and C.114. Apart from the fact that the fuel tap is not in the instrument itself and that the actual layout of the carburettor is different, adjustment of the Keihin PW.16 instrument fitted to these models should be carried out as described in the sections relating to the Model C.100. The throttle screw, which has a knurled head, is set immediately above the pivot point for the float chamber clip and the air adjuster screw is set at an angle to its left.

Carburettor Cleaning: Models C.110 and C.114. Detach the mixing chamber top and draw out the throttle slide assembly. Tape the assembly out of harm's way. Detach the oil feed unions (which carry warm oil to the choke area), the fuel pipe, and unbolt the carburettor from the frame. Slide the instrument off its rubber hose and wash it in clean petrol. Then dry it with non-fluffy rag. Spread out clean newspaper to work on, and have a small bowl of petrol at hand for washing purposes.

Prise off the spring clip holding the float chamber to the body. Take out the hinge pin on which the twin floats pivot, and place them on one side. The needle valve can then be removed for cleaning. Projecting down from the body is the hexagonal needle jet holder with the main jet in its base. Remove both for cleaning. When the holder is out the needle jet itself can be carefully pushed downwards and out of the carburettor body. Clean each jet by blowing through it from the far end and wash it in petrol. Then replace them and take out the pilot jet which will be found just to the side of the main jet. Clean this and then replace it. Wash the floats and float chamber. Examine the floats for any signs of damage. To test for leakage, shake the float assembly. If petrol has entered you will hear it swill around inside.

Rebuild the carburettor using new gaskets throughout. Make sure that the float needle is fitted properly, with the taper end bearing on the valve seating. To test for the correct petrol level invert the carburettor with the float chamber bowl removed. The float needle spring will then be depressed by the weight of the floats. Carefully tilt the carburettor until it is between fifty and seventy degrees out of the vertical and measure the distance between the carburettor body base and the outer periphery of the float. It should be exactly 19·5 mm, though it suffices if the distance lies between 19 mm and 20 mm. To alter the petrol level bend the float tongue very gently, and recheck the measurement. But be particularly careful not to alter the position of the individual floats in doing so.

Carburettor Adjustment: Models P.50 and PC.50. As with other Hondas, the carburettor on these models rarely needs attention. The standard adjustment is that the throttle needle should be set on its third notch, although for specific conditions it is permissible to vary this by one notch either way. In general, follow the instructions given earlier in this chapter under the heading for the Models C.100 and C.102.

All other adjustment is external, and should be made with the unit at normal running temperature. First, the air screw should be slowly rotated until it is fully home on its seating. Don't over-tighten it, or the face will be damaged. Then back it off by exactly $1\frac{1}{8}$ turns (up to $1\frac{3}{8}$ turns on the PC.50). Next, loosen the throttle stop screw fully, before screwing it in to give a good reliable idle—1500 r.p.m. You can check this by running the engine with the rear wheel clear of the ground. The correct idling setting is the maximum you can obtain on the throttle stop screw before the rear wheel starts to turn.

If the engine is still not running smoothly, you can now make further minor adjustments to the air screw—no more than a quarter-turn either way—and compensate for faster or slower running on the throttle screw.

Air Cleaners: All Models. Paper-type air filters are used on most 50 c.c. Hondas. On the C.100, C.102 and C.50 the filter is mounted on the frame tube just behind the steering head, and on the small motor-cycles it is on the right side of the machine beneath a plastic cover. On the P.50 and PC.50 it is below the saddle, on the left hand side.

In the first case, unscrewing one domed nut and lifting an alloy cover bares the filter, and removal of a second nut enables it to be lifted out. In the motor-cycles, remove the cover by undoing the knurled screw and then take the filter from its place. On the mopeds, remove the cap and pull out the element and its set plates.

Clean the filter and the housing by dusting it with a soft brush. You can also use a tyre pump to blow away surface dirt. Make sure that the element is not damaged, damp, or fouled by oil, then refit it.

Cleaning Fuel Tap: Models C.110 and C.114. The filter bowl is screwed to the bottom of the fuel tap and can be removed by use of a 10 mm ring spanner or socket. Do *not* use an open-ended spanner. It may also be necessary to support the tap with an appropriate-sized spanner while undoing the bowl. Lift out the synthetic rubber sealing washer and the filter screen. Wash the screen and the bowl in petrol, and then refit them.

Brake Adjustment: All Models. Screw-type adjusters are provided for all brakes. Take up wear by turning the adjusters half a turn at a time. Except on the front brakes of the P.50 and PC.50 models they have a "click" action. Continue adjustment until the brake is felt to bind when the wheel is turned. Then slacken off, half a turn at a time, until binding disappears. Check the efficiency of the brakes regularly.

Wheel Removal: All Models. To detach the front wheel place a suitable wooden block under the crankcase so that the wheel is clear of the ground. Disconnect the brake adjuster nut and take out the nut and bolt which, on the C.110 and C.114 variants, secures the brake plate to the anchorage arm. Disconnect the speedometer at the drive end, take out the front axle split pin, and undo the spindle nut. The axle can now be pulled out of the hollow wheel spindle, and the wheel will be displaced from the two fork links. Finally free it by removing the brake cable from its housing on the backplate.

To take out the rear wheel disconnect the rear brake rod adjusting nut and take off the nut and spring clip securing the brake anchor arm. Remove the 17 mm nut on the rear axle and pull the axle out. The tubular distance piece on the right side between the brake and the fork can now be lifted away and the wheel is simply pulled to the right to free it from the drive and then dropped out.

On the P.50, follow the procedure given in chapter 10. On the PC.50, note that the chain cover must be removed, the chain disconnected at the spring link, the brake cable freed at the arm and the chain adjusters loosened. The axle nut can then be freed and the wheel pushed *forwards* to free it.

Early C.50s did not have a quickly detachable wheel. On these, the chain case has to be removed, the chain link disconnected, the brake adjuster undone, and the torque arm freed before the wheel nuts are released and the wheel removed.

Silencer: Models C.110 and C.114. Cleaning the baffle tube of the silencer is recommended every 2,500 miles. To reach it, detach the 5 mm screw which is to be found at the end of the silencer and draw out the baffle tube. Unblock all holes and brush away all carbon deposits.

Bulbs: All Models. To reach the headlamp, speedometer, and neutral-indicator bulbs remove the front of the headlamp, which is held by a recess-headed screw. To reach indicator and rear lamp bulbs take off the appropriate lens assembly by removing the recess-headed screws.

Stoplight Timing. The stoplight switch in each case is operated from the brake pedal through the medium of a spring. If the light is lighting too late—or is not lighting at all—it indicates that greater tension is required. Loosen the lower switch lock nut and lift the switch by screwing the top nut downwards. This advances the operation of the stoplight. When the light is coming on at the desired amount of pedal movement retighten the lower nut. To make it light later, reverse this procedure.

Chain Adjustment: All Models. Driving chains are adjusted by pulling the rear wheel backwards by means of drawbolts. To do so loosen the rear axle nut, and then the nuts on the drawbolts should be turned one flat at a time to tension the chain. After each nut has been moved by this amount check the up-and-down play on the lower chain run through the hole in the chain case, normally closed by a push-in plug. Total movement must not be less than 0·4 nor more than 0·8 of an inch. When the correct play has been achieved, retighten the axle nut. Finally check that the guide marks on the left and right drawbolts are in the same relative positions. There are corresponding marks on the fork members. If the marks on both sides do not coincide the drawbolt setting must be altered so that the wheel comes into correct alignment. Naturally, the chain tension will then have to be rechecked. Always test tension with the machine off the stand and with a rider aboard.

The rear chains on some models tend to require frequent adjustment as a result of chain stretch. This can be minimized, I have found, by giving regular attention to chain lubrication.

Every 500 miles, lubricate the chain with Brush-on Linklyfe. This ensures that the drive does not run dry. At intervals of 2,000 miles I have found it advisable to detach the chain and wash it in petrol, following this with an immersion in solid Linklyfe melted over an electric or gas ring.

Battery: All Models. Batteries require little care—but they demand it regularly. The battery loses part of the water component in its acid when it is in use. This loss must be made good by adding *distilled water* to bring the electrolyte level back to the upper level mark on the battery casing. This water is added by detaching the battery from the machine —it is held by a clamp which has a single locking bolt, and by its two leads which are bolted to it. Then take out the cell plugs and, using a suction-type filler, add the distilled water to each cell in turn. Clean the terminal posts and lead terminals with emery cloth until the metal is bright, and coat them with Vaseline before refitting the battery.

Rear Suspension Lubrication: Models C.100 and C.102. On the C.100 and C.102 scooterettes, each rear suspension unit has a greasing plug set in the upper lug casting. Periodically—once annually should be ample—these plugs should be removed and 1 c.c. of light grease injected into the interior of each suspension leg. Do not exceed this quantity.

9 Engine overhauls, pushrod o.h.v. models

As one of the main objects of home maintenance is to save money, there is very little point in attempting to strip one of the 50 c.c. Honda power units completely. These engines are engineered, of necessity, to extremely fine limits. Therefore, an extensive kit of special tools is needed for work of this nature. Since such a kit can cost upwards of £20 its acquisition is not economic, and it is better to entrust the very rare complete overhaul to an agent's well-equipped workshop. However, for the benefit of enthusiasts who wish to know what is involved I have set out the basic stripping procedure at the end of this chapter.

Most of us, I think, are more likely to wish to save the odd pound on top overhauls—a job which is well within the average owner's compass. This can be done with the engine still mounted in the frame, too, which is a distinct advantage for those who lack workshop facilities.

Detach the front shielding by releasing its four securing bolts, two clamping screws, and the single centre nut holding the air filter cover. Lift the shield from the machine. Then, using a paintbrush, coat the entire engine unit with grease solvent. Work this into the cylinder fins, and into all the odd corners. Next, wash the solvent away with water trickled on from a garden hose, and then give the engine time to dry off. I normally let mine stand overnight.

STRIPPING: C.100 AND C.102

When the components are dry stripping can begin—but before you do this make certain you have all the tools, materials, and spares which you need. They are a decoke set comprising new gaskets for the rocker cover and cylinder head joint, new rubber oil seals for the oil drain and pushrod tunnels, a copper/asbestos ring for the exhaust port, and new valve springs. Fine grinding paste is another necessity, together with some clean petrol and a suitable bowl so that components can be washed. For clearing carbon a scraper will be wanted, and for removing the valves a small valve-spring compressor.

As a first step take off the auxiliaries—the carburettor, the dirt shield, and the exhaust pipe. Detach the plug cap and take out the plug. All this is straightforward. The carburettor merely demands removal of the mixing chamber top and slide and the fuel pipe. But remember that the tank must be drained, or some suitable plug made up ready to prevent the fuel from flowing out of the pipe since the tap is on the carburettor. Take off the air hose. Only two 10 mm nuts then hold the carburettor to the barrel.

The dirt shield is held by four 10 mm nuts and the exhaust pipe by two 10 mm nuts at the cylinder head and two 10 mm set screws at the silencer support bracket.

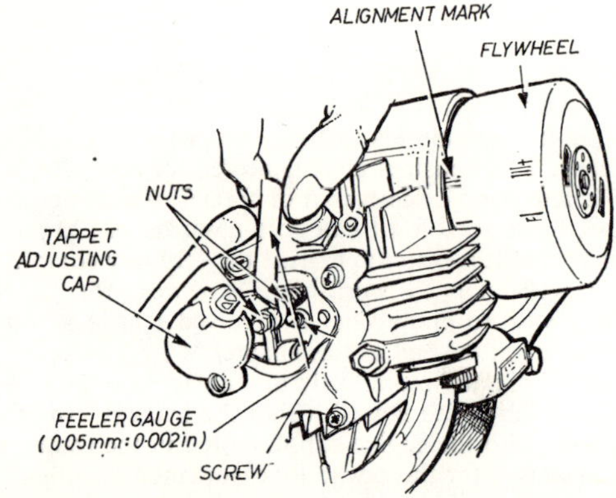

Fig. 18. Tappet adjustment method. How the job is done. The flywheel marks are correctly aligned, and the gaps measured with a feeler gauge

STRIPPING: C.110 AND C.114

Follow the above sequence, with the exception of the carburettor and dirt shield. For the carburettor, follow the instructions in the previous chapter.

Head Removal. Now take off the rocker feed oil pipe, using a 10 mm spanner on its two banjo unions, and the two rocker cover inspection caps, each of which has a 17 mm hexagon. Take off the four 10 mm bolts which hold the rocker box and lift this away from the cylinder head. Spin the pushrods round to free them, and then pull them out.

A further four nuts hold the actual cylinder head. Remove all the nuts, and slide the head off its studs. If the joint is difficult to break, cover the head with a pad of cloth and jar it with the heel of your hand. *Never* try to insert a screwdriver into the head joint or the head and barrel may be irreparably damaged.

Barrel Removal. Once the head has been taken off, removing the barrel is merely a matter of sliding the cylinder off its studs. Support the piston

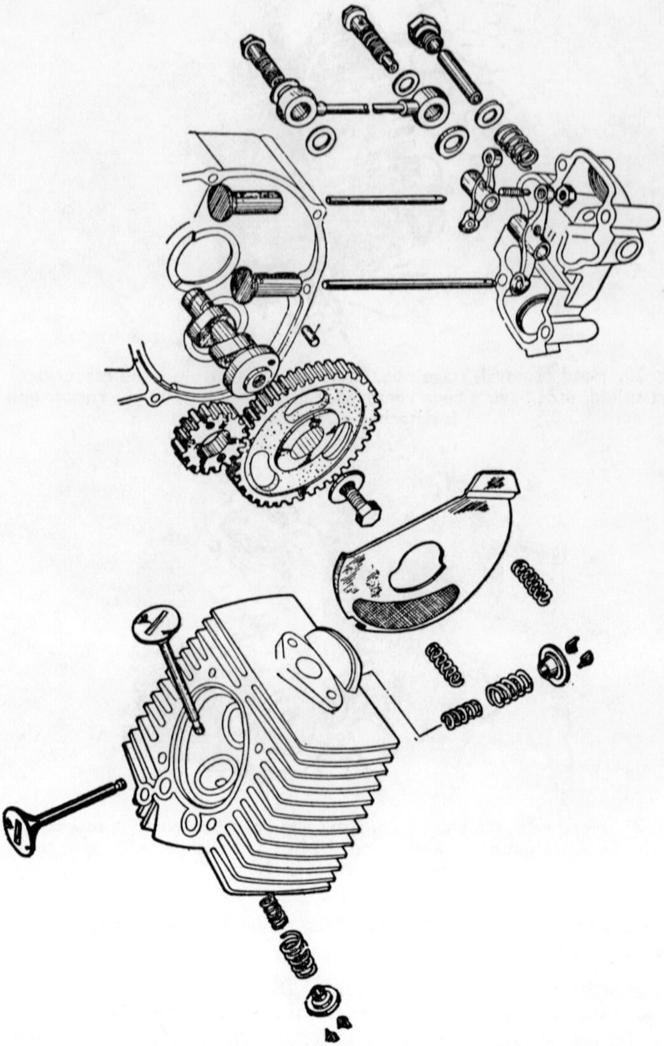

Fig. 19. Valve gear, C.100, C.114. This exploded view shows clearly the layout of the valve gear. For decarbonizing, the valves must be removed from the head and their seats must also be checked for signs of pitting or burning

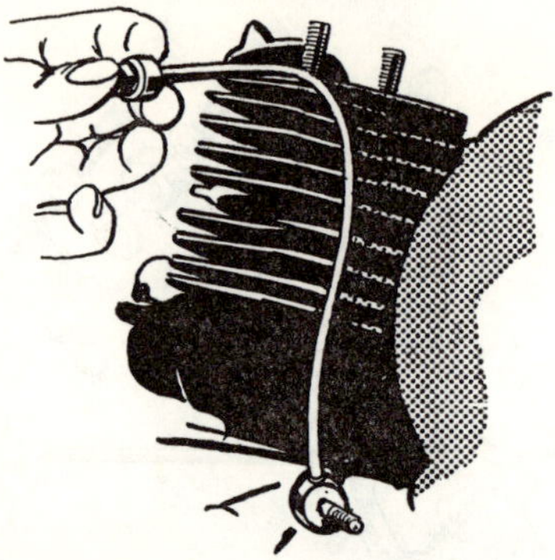

Fig. 20. Head removal, stage one. First, the exhaust pipe, the carburettor dirt shield, etc., having been removed, the oil feed pipe to the rocker box is detached at both ends

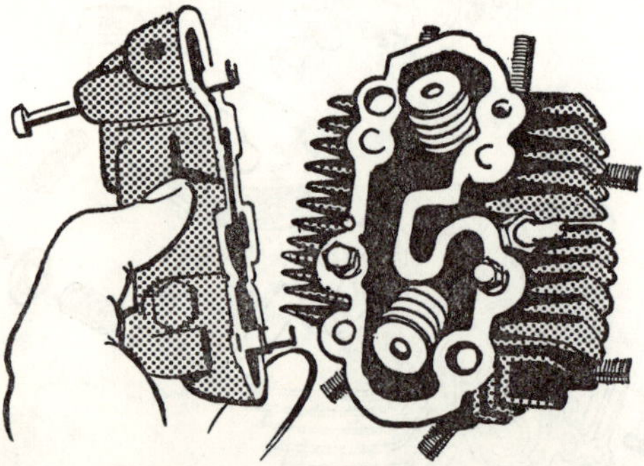

Fig. 21. Head removal, stage two. Next the bolts securing the rocker box to the head are undone. Then lift off the box and withdraw the two push-rods from their tunnels

as it emerges from the lower end of the cylinder, or it may fall against the engine and be damaged.

Piston Removal. Two circlips position the gudgeon pin in the piston boss. Use a pointed instrument to release these and then gently tap the pin out of the boss. If it is hard to move soak rags in hot water, wring them out, then wrap them round the piston and leave them for a minute or two. This will expand the metal slightly and make the pin easier to move.

Fig. 22. Head removal, stage three. Now the four head nuts are taken off and the head is free to come off its studs. Should it stick, jar it with the heel of your hand

Fig. 23. Barrel removal. With the head off the barrel can slide from the studs. Support the piston as it leaves the bore, to prevent it being damaged

Valve Removal. Slide the valve-removing tool over the valve head, with its feet on the valve cap. Screw the tool down until the springs have been compressed enough to free the collets which secure the valve stem in the cap. Remove the collets and release the tool. The valve, spring, and cap can then be lifted out. Mark the head of each valve to identify it.

Carbon Removal. With the scraper, clear all carbon away from the inside of the cylinder head and from the exhaust port interior. Make sure that the area around the plug hole is clear. Finish off by using wire wool to give a good surface finish.

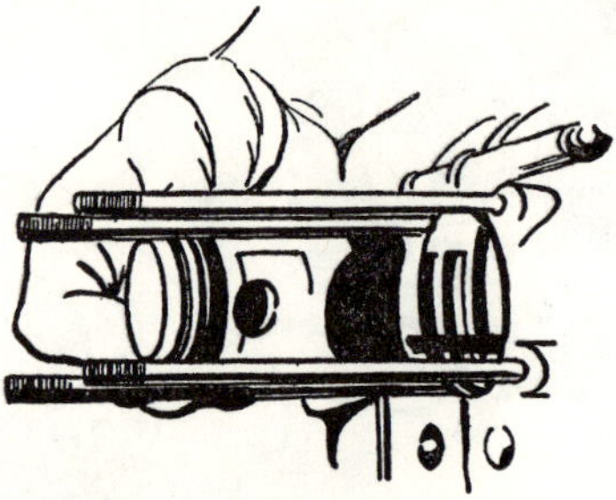

Fig. 24. Piston removal. Detach the circlips at each end of the gudgeon pin. Warm the piston with rag well wrung out in hot water. This expands the light alloy. The gudgeon pin can then be pressed out of the piston bosses easily

Fig. 25. Piston fitting. When refitting the piston to the connecting rod remember that the arrow stamped on its crown must point downwards. After fitting the piston replace the barrel. Do not omit the rubber sealing rings (arrowed) on the pushrod tunnels and oilway

Similarly clean both valves, including that section of each valve stem which is exposed inside the ports. Do not scrape or polish the area which is already bright. This is the working surface inside the valve guide.

Gently lift out the piston rings by expanding them slightly with the thumbs. Scrape the piston crown clear of carbon, and finish off with wire wool. You may, if you wish, polish both the inside of the combustion chamber and the piston crown with metal polish, but do *not* use emery cloth. Check that the ring grooves are clear of carbon, and that there is no carbon behind the rings themselves, If there is, use a piece of broken ring as a groove scraper and a blunt penknife to clean the rings. Under no circumstances must the outer surfaces of the rings be touched.

Valve Seating. Honda advise refacing any valve on which pit marks do not exceed a depth of 0·006/0·007 of an inch. Any which have deeper pitting than this should be renewed completely. For the slight cost involved, it is better to use new valves. At a pinch, however, re-cutting will suffice. You will need to get a Honda agent to face the valves, on a machine,

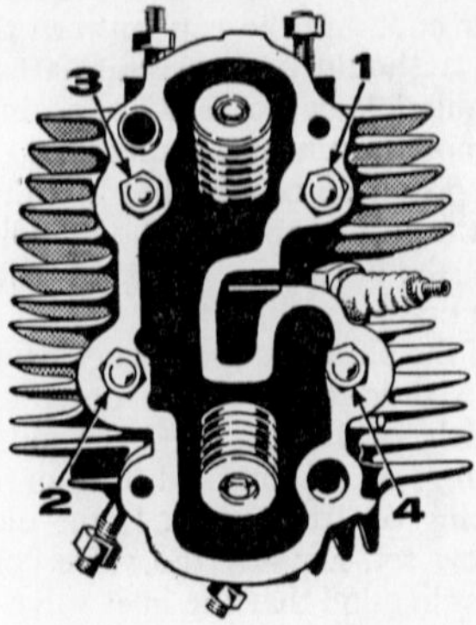

Fig. 26. Cylinder head nut tightening. It is essential that the head should be drawn down evenly by tightening the nuts gradually, working from one to another in the order shown here. If a torque wrench is available, use it. The correct figure is 70 in./lb

to an angle of forty-five degrees. Pitted seats in the head should be similarly recut.

Before fitting, both valves should be lightly ground with a fine grade of paste. Smear the valve face with a thin layer of paste and drop the valve into position. With a broad-bladed screwdriver inserted into the slot in the valve head work the valve from side to side through an arc of about ninety degrees, quickly, about half a dozen times. Then lift the valve, give it a quarter turn, and repeat the process. Carry on with this sequence until both valve and seat have a thin but continuous grey line around them. The grinding is then complete.

At this stage, remove the valves and *wash them and the head thoroughly in clean petrol.* Every trace of the abrasive grinding paste must be removed before assembly or it may enter the engine and cause very serious wear. Also wash the piston if metal polish is used on the crown.

Reassembly. Rebuilding the engine is virtually a simple reversal of the stripping procedure, but there are one or two points to watch. You will find that the piston rings carry maker's marks. These go upwards. The top ring is marked with the word "Top" and obviously this mark must be uppermost. The top ring is a chrome type and the lower ring a slotted scraper design. Fit this from the skirt end of the piston. Gently ease the middle ring into the centre groove by working from the top, and finally add the top ring.

If the existing piston rings have been reused there will be no need to check the gaps. Where new ones are being fitted, however, they should first be inserted into the bore by themselves, squared up by sliding in the piston from the other end, and the gap between the ring ends measured with a feeler gauge. It should be between 0·003 and 0·010 of an inch, with a maximum permissible gap of 0·039 of an inch. Gaps are ready set on Honda rings and no adjustment is required.

When refitting the piston be certain that it is the right way round. An arrow is stamped on the crown, and this should point downwards. Lightly oil the bore and slide the barrel on. Compress each ring in turn so that it slips easily into the bore.

Check the length of the old valve springs against the figures given in the appendix. Permissible shortening of the outer springs is 1·4 mm on C.100 and C.102 models, or 0·6 mm on C.110 and C.114 machines. With inner springs shortenings of 1·3 mm and 0·9 mm respectively are permissible, both figures being on free length. If the old springs have passed these limits, use a new set. Employ the valve compressor tool to refit each valve in turn. Bear in mind that the inlet valve is slightly longer than the exhaust valve. Don't mix them up.

Refit the head, using a new gasket, and tighten the head nuts finger-tight at first. Then, using a socket spanner, work from nut to nut, diagonally, taking each up a few threads at a time until the head is securely seated.

Insert the pushrods—the exhaust pushrod is shorter than the inlet one —and replace the rocker cover. Use a new gasket, lightly oiled. An old gasket, if re-used, must be annealed by raising it to red heat, then quenching it in a bath of clean water. Tighten the rocker bolts and reset the tappet clearances. Then complete the assembly work, before attempting to test the engine.

Complete Dismantling. This abridged section is intended simply for the guidance of owners with workshop experience and the requisite set of Honda service tools. I do not advise the average owner to attempt work of this nature.

First remove the engine from the frame by disconnecting the fuel lead, the electrical wiring at the snap connector points, the control cables (to the throttle in the case of the scooterettes, to throttle and clutch on C.110 and C.114 models), and the chain. This latter entails removal of the lower chain case half, the short chain case top cover, and the side cover of the engine on the drive side. The oil must be drained and the battery leads removed.

Detach the carburettor, the footrest assembly the starter motor feed cables on C.102 models, and unhook the brake and stop-lamp switch springs. Remove the exhaust system. Place a block under the engine unit, take out the upper and lower mounting bolts and lift the engine unit from the frame. On C.110 and C.114 machines the gearchange pedal also has to come off before removing the engine.

Once the unit has been cleaned and is on the bench use the special holding and withdrawal tools to remove the flywheel or, in the case of the C.102, the rotor, and then the stator plate. Lock the final drive sprocket,

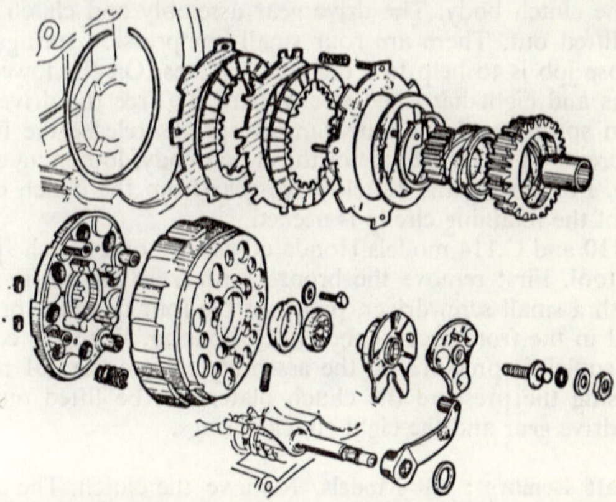

Fig. 27. The Honda automatic clutch design. Employed on the C.100, C.102, C.105E and "Monkey Bike" models this clutch is operated by centrifugal force, yet also incorporates a two-way positive control to ensure easy gear changing and kick starting

and remove the two bolts which hold it. Take out the neutral indicator switch, which is held in place by a single recess-headed screw.

Follow the sequence outlined earlier in this chapter for stripping off the head, barrel, and piston.

Nine screws hold the primary drive side cover, removal of which gives access to the clutch. A special tool is required for the clutch nut. As there is a pair of small compression springs inside the cover—their job is to retain the oil plate in front of the camshaft drive gear—the cover will tend to spring off. Take off the plate, and the anti-rattle spring fitted between the clutch operating cam plate and the release mechanism. Remove the clutch operating lever and washer (from the splined shaft) and the clutch operating cam plate can be lifted away. Then prise the journal ball bearing and carrier from the centre of the clutch drum.

On the C.110 and C.114 the actual clutch dismantling is similar, but the operating pushrod has a lubrication reservoir. An arm is attached to this trough at one end and to the centre of the primary gears shaft at the other. Remove this, and the ball journal and carrier.

Clutch Removal. With the clutch body securely held, free the centre nut's tab washer and release the nut. Slide the clutch off its shaft. To dismantle the clutch unit of C.100 and C.102 models take out the bronze bush from the drive gear centre and, with the drive gear facing upwards, prise out the 101 mm diameter spring ring which is set in a groove in the rear of the clutch body. The drive gear assembly and clutch plates can then be lifted out. There are four small compression springs, on guide pins, whose job is to help free the clutch plates. On the lower plate are guide pins and eight hardened steel rollers. To free the drive plate, the four main springs and the four damper springs release the four cross-headed screws on the front face of the clutch body, loosening each a little at a time. For freeing the clutch drive gear from the clutch centre only removal of the retaining circlip is needed.

On C.110 and C.114 models Honda advise use of a clutch spring compressing tool. First remove the bronze bush from the centre drive gear. Then, with a small screwdriver, prise out the four damper springs which are found in the front face of the clutch housing. With the compressing tool put sufficient pressure on the assembly to free the 101 mm circlip. On releasing the pressure the clutch plates can be lifted out, together with the drive gear and the eight clutch springs.

Camshaft Removal: All Models. Remove the clutch. The camshaft—complete with the gear—can now be taken off, followed by the cam followers. To remove the camshaft timing pinion from the crankshaft a special puller must be used.

Splitting the Crankcase. Remove the 17 mm circlip from the primary shaft of the gearbox and draw off the driver gear. Take the 23 mm circlip and washer from the kickstarter shaft and disconnect and remove the starter spring. A pair of pointed-nose pliers is needed here. Remove the gearshift operating arm and small stopper arm. Set the crankcase down on the generator side, with the nine crankcase bolts removed, and lift off the righthand half. The crankshaft assembly, kickstart, and the entire gear cluster will remain in the generator side crankcase half.

Retiming the Valves. To time the valves on early models, drive the timing pinion on to the mainshaft using the special tool and mesh the crankshaft pinion tooth marked with a dot into the correspondingly-marked sector of the camshaft wheel.

Later models have no special marks on the camshaft pinion. Instead, the tooth immediately above the extractor hole—the one which is five teeth round from the keyway—is used instead.

Make sure which system applies to your model before you rebuild the engine.

STRIPPING PC.50/PF.50, O.H.V. VERSION

In general, the o.h.v. variation of the PC.50 engine—also used on the PF.50—is not significantly different in layout from the o.h.c. original. The major innovation is, of course, in the substitution of a push-rod valve operating mechanism for the o.h.c. chain drive, and accommodating the camshaft in the crankcase instead of placing it in the cylinder head.

So far as top overhauls are concerned, owners of o.h.v. models can follow the instructions given for the earlier version in the following chapter, with the following exceptions:

Detaching the Cylinder Head. Remove the valve cover and release the rocker lock nuts, holding the adjuster nuts stationary with a spanner as you do so. Then undo the adjuster nuts, lift out the rockers, twist the pushrods smartly, and pull them away from the engine. Undo the four cylinder-head-retaining bolts, and ease the head from the barrel.

Engine Reassembly. Fit the valves, springs and retainers. Then assemble the rockers, complete with their adjuster nuts and locknuts. Insert the push-rods in their aperture in the barrel, fit a new cylinder head gasket, and offer up the head. Feed the pushrods into position as you do so.

Obviously, if the camshaft position is such that one or both valves would be open, the head would not seat snugly. Before fitting the head, therefore, press lightly on the ends of the rods and turn the engine so that it is set at Top Dead Centre on the power stroke—when both rods will be at their minimum travel. Once the head is seated, insert the securing bolts, and tighten them in sequence. Reset the tappets to 0·002 in. cold.

Stripping the PC.50/PF.50 o.h.v. units. It is unlikely that the average owner would wish to carry out a full engine/transmission strip. The method is, however, given here in outline:

1. With the engine on the bench, remove the generator.
2. Detach the cylinder head, valve gear, and barrel.
3. Remove one piston pin circlip and press the pin out to release the piston.
4. Detach the right-hand crankcase cover.
5. Bend back the clutch retaining nut's lock washer and, with the clutch firmly locked, undo the retaining nut. Detach the centrifugal shoe assembly and the clutch outer drum.

6. Remove the driven gear, and then lift out the tensioner spring.
7. Take out the timing sprocket, camshaft and drive chain as one unit.
8. Undo the screws securing the crankcase halves and split the cases.
9. Lift out the crankshaft.
10. There are inner and outer 20 mm washers and circlips on the pedal shaft. Remove these, and detach the drive sprocket from the shaft, releasing the pawls and springs at the same time.
11. Detach the main transmission shaft.
12. Slide the cam followers out of the crankcase.

Reassembly Sequence. Start reassembly by fitting the starter chain sprocket, complete with pawls and springs, to the pedal shaft, securing it with the circlip. Then replace the cam followers. Fit the starter sprocket, pedal axle and chain; the main transmission shaft; and the crankshaft.

After rebuilding the crankcases, offer up the camshaft. To set the valve timing, turn the crankshaft to the Top Dead Centre position. Fit the drive sprocket and camshaft sprocket to the chain so that their timing marks face each other, and offer them up as a unit. Note that it will be impossible to refit the camshaft unless the cam followers are first lifted clear of the inner face of the crankcase. Slide the distance collar on to the crankshaft, with its smaller diameter next to the sprocket, and then fit the clutch guide bush.

Replace the clutch, making sure that the three steel balls are located in the ball retainer plate, and that they are set in the lowest point of the tracks in the clutch outer unit. Then offer up the clutch assembly, and after tightening its retaining nut peen over the lock washer.

Refit the piston, barrel, cylinder head, generator, and crankcase cover.

10 Engine overhauls, o.h.c. models

WITH the introduction of the C.50, P.50 and PC.50 models, Honda extended the use of ultra-efficient overhead camshaft valve layouts to their entire range. The C.50 is the successor to the well-tried o.h.v. C.100 series, and—like its predecessor—has a three-speed gearbox. Both the other machines are single-speeders, the P.50 having an "engine in the wheel" layout and the PC.50 an under-bracket engine mounting.

OVERHAULING THE P.50 POWER UNIT

Before starting work on the engine, warm it up slightly and then drain the oil. If this is not done there is a danger that, as the unit is removed from the machine, oil will escape into the brake housing—with predictable and wholly unwanted results!

Removing the Engine. Once the oil has drained, switch off the petrol at the fuel tap. Then remove the carburettor shield. This is a spring fit in the frame members.

Detach the H.T. lead from the plug and disconnect the decompressor cable at the engine. Free the choke cable at the carburettor, loosen the lower clamp of the air cleaner hose, ease the fuel pipe from its stub, free the carburettor clamp, and remove the carburettor from its stub.

If you wish to work on the instrument, unscrew the mixing chamber top and pull out the throttle slide. Otherwise, tape the complete carburettor out of the way, behind the saddle tube.

Now detach the rear brake cable at the wing nut, and undo the electrical leads from the engine unit at their snap connectors. Remove the domed nut securing the rear torque arm to the engine and the 8 mm bolt through the left upper chain stay that holds the front of the arm.

Undo the two nuts below the cylinder head that hold the exhaust pipe and the single bolt on the other side of the machine securing the silencer to the frame. Lift away the exhaust pipe. Now loosen the rear axle nuts sufficiently to enable the entire rear wheel assembly to come free from the frame. To remove it, lift the frame upwards, away from the wheel, slipping the pedalling chain from its sprocket once the wheel assembly is free. Alternatively, you can disconnect the drive chain at the spring link before removing the wheel—a procedure that is certainly less "fiddly."

Detaching the Cylinder Head. Remove the rocker cover. It is held by four cross-headed screws. It comes away complete with the rockers, rocker shaft and decompressor mechanism, and there will normally be no need to separate these. If there is actual damage to the rockers, or wear on the mechanism, the shaft should be extracted through the aperture in the right-hand side of the cover, normally closed by a plug with a cruciformed screwdriver recess.

Next, pull out the locking pin from the recess in the camshaft bearing boss just below the exhaust valve and screw a bolt into the camshaft centre pin so that this may be withdrawn. When this is out the drive chain can be eased off the camshaft sprocket.

Remove the four 6 mm nuts and single 6 mm screw that hold the head to the cylinder, and slide it off its studs.

Valve Removal. Shape a small piece of wood to fit neatly inside the combustion chamber, and place the head against the bench so that the valves cannot move. Press each valve retaining cap in turn, until the recess in the cap coincides with the groove on the valve stem. Press the cap sideways, so that it is free of the groove, and lift it off followed by the spring. When

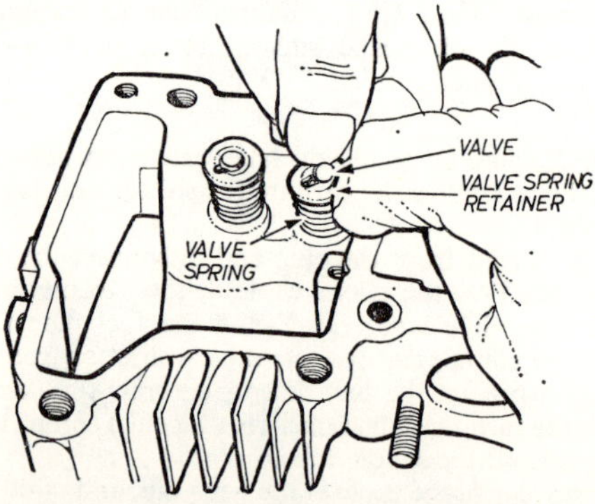

Fig. 28. Valve removal, P.50/PC.50. On these machines, the valve caps are slotted to facilitate removal and fitting

both springs are off, remove the wooden former from the combustion chamber and slide the valves out of their guides. Do not mix them up if you plan to re-use them. There is, however, a good case for renewing both the valves and the springs whenever the engine is decarbonized. The cost is small. Early P.50s had a different type of exhaust valve that lacked the waist below the head that is used on current models. Where the earlier type of valve is still fitted, it should be discarded and the later pattern used as a matter of course. New-type retainers should also be fitted.

Decarbonizing and Valve Grinding. In general, follow the procedure set out on pages 65–7. If, however, you are having the head refaced, it is advisable to ask your dealer to recut the valve seats and fit the new valves, retainers and springs at the same time.

If you carry out the regrinding yourself, finish up with a petrol seepage test. Fill the combustion chamber with petrol, and leave it for a few minutes. Note if any has seeped through into the induction or exhaust tracts. If so, the valves are leaking and further grinding will be required. Otherwise, there will be a loss of efficiency in the engine.

Valve Guides. Although it is possible to remove and refit the guides yourself, it is necessary to ream them to precise dimensions afterwards. Again, this is a job better left to a dealer.

The exhaust valve guide must be reamed to an internal diameter of 0·1898–0·1902 in. (4·82–4·832 mm) and the inlet to 0·189–0·1894 in. (4·8–4·812 mm). The upper edges of both guides must be exactly level with the upper (rocker cover) face of the cylinder head. The valves themselves have standard diameters of 0·189–0·188 in. (4·79–4·78 mm), and any valve that has a stem diameter below 0·187 in. (4·74 mm) must be renewed.

Engine Reassembly. When the valves have been replaced and the sealing has been proved satisfactory, fit a new head gasket. Be careful, too, not to omit the two hollow dowel pins that fit on studs Nos. 1 and 2. Offer up the head, carefully feeding the cam chain into the tunnel, and fit the head nuts and the cruciform-headed screw. Torque the nuts to 6·5–8·7 lb fit (90–120 kg cm), following the sequence shown in Fig. 30, and working a few threads at a time from one to another. This point is important. If the nuts are not taken up evenly, or the torque is too great, the head joint will be distorted and compression will be lost.

Now remove the flywheel cover, and—keeping the cam chain pulled tight—turn the flywheel until the T.D.C. mark (T) is against the index mark cast on the crankcase. Slide the camshaft into place so that the twin holes in the sprocket are parallel to the rocker cover face of the cylinder head. Later-type sprockets carry a stamped arrow head and have no holes. With these, the arrow must be at the top, and aligned with the axis of the engine.

Next, check the O-ring on the camshaft centre pin, and insert the pin carefully, so that the ring is not damaged. The holes in the pin and in the oil guide alongside the cam sprocket *must* be aligned as the pin is tapped into place. Then align the lock pin holes in the camshaft and the cylinder head, and drive in the dowel pin on the far side.

Refit the rocker cover, with the separate tappet cover removed, and turn the engine over several times to ensure that it is operating smoothly and that the valve timing is correct. Reset the tappet clearance to 0·002 in., cold, on both valves and then fit the tappet cover.

Ignition Timing. Remove the screws that hold the flywheel-generator cover. Then turn the flywheel until the "F" mark stamped on its circumference is in line with the indicator mark on the crankcase, just behind the upmost part of the cylinder barrel/crankcase joint. At this setting, the contact-breaker points should be just opening—a tiny gap of 0·0015 in., which is about the thickness of a cigarette paper. You cannot measure this with any accuracy, but if you use one of the cheaply-bought circuit test lamps (you can even make your own) the job is easy.

The lamp must have its own battery and two leads terminating in crocodile clips. The battery should be permanently in circuit—i.e., touch the two clips together and the lamp lights. Arrange the bulb so that it is in a separate lead joining the two crocodile clips.

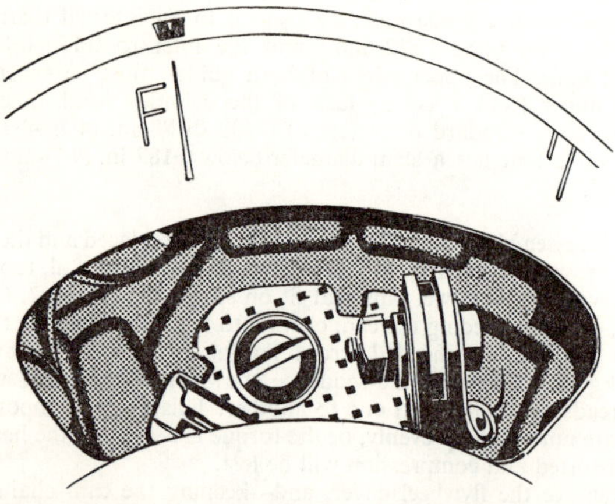

Fig. 29. Contact points and timing, ALL models. With the "F" mark on the flywheel aligned with the line on the crankcase, the contact points should just be opening. This is the only timing adjustment possible on the Hondas, fitted with flywheel generators

With this tester, simply attach one clip to earth on the engine. Then undo the black lead from the generator to the engine at the connector cluster, and attach the other clip to the end of the generator lead. With the points closed the lamp lights. Turn the flywheel until the lamp dims or extinguishes —which it does as the points open. Note the position of the "F" mark. If it has not reached the crankcase mark the points are opening early and the ignition is over-advanced. If it has passed it, they are opening late and it is retarded.

Adjust it by setting tiie two marks together. Then loosen the contact breaker lock screw, insert a screwdriver into the adjuster slot on the plate, and twist it until the lamp lights. Then gently turn the screwdriver until the lamp shows the points are open. Hold this adjustment while the plate locking screw is tightened, and recheck the timing with the test lamp in case the setting altered as you locked the screw. Then reconnect the black lead.

Contact-breaker Points Setting. When fully open the points gap should normally be 0·012 in–0·016 in. However, the actual gap is immaterial providing the points are breaking at the correct moment, as described in the preceding section.

Where the test-lamp procedure is impracticable—as it might be, for example, at the roadside, you can as a temporary measure set the points fully open and gap them accordingly.

If this is done, the correct timing procedure should be carried out as soon as possible afterwards.

Contact-breaker Points Replacement. Worn or dirty points will rob you of power. To gain access to the contact-breaker, the flywheel rotor must be removed—for which a puller is required. Simply remove the centre nut, screw in the puller, and hold the flywheel steady while the puller's centre screw is tightened down to draw the flywheel off its taper. Take care not to lose the Woodruff key that locates the flywheel on the shaft.

The moving arm of the contact breaker pivots on a post fixed below the plate that carries the static point, while the spring is bolted to the plate at its free end. There are insulating washers here, and if the moving arm is removed it is essential that the sequence of the washers is carefully noted and that they are replaced in their original order. Otherwise, the ignition system will be short-circuited.

You will still need to note the washer positions even if you intend fitting a brand-new set of points, since the black generator lead also connects to this bolt, and its tab must be correctly placed in relation to the insulators.

To remove the contact-breaker system, simply slide the circlip on the pivot post out of position and loosen the terminal bolt on the fixed plate. The moving contact arm can then be slipped off its pivot. Then free the generator lead from the bolt, undo the locking screw, and take off the fixed plate. If the contact-breaker points are not badly out of square or deeply pitted they can be squared-up on a stone, refitted, and the gap set as already described. If there is any doubt of their condition, however, it is better to renew them. Remember to wash off any preservative on the faces of the new points.

Renewing Piston Rings. To obtain access to the rings, follow the dismantling procedure already detailed as far as removal of the cylinder head. Then set the piston so that it is at the open end of the barrel, and carefully pull the cylinder off its studs. Support the piston as it clears the mouth of the barrel, so that it does not sustain damage.

The piston is held to the connecting rod by a gudgeon pin, which is located by circlips. Remove these by compressing their projecting ends with a pair of long-nosed pliers and gently pulling them away from their grooves in the piston. Then press out the pin. If it is stiff, warm the piston by wrapping it for about a minute in clean rag wrung out in hot water. This will expand the metal and free the pin.

The rings can then be sprung from their grooves for cleaning. Remember that they are brittle and snap easily. It therefore pays to have a spare set handy, just in case of accidents. Note, also, that the rings are not interchangeable. Each *must* go back in the groove from which it was removed.

Carefully scrape all carbon from the upper and lower surfaces of the rings, and also from their inner edges. Do NOT scrape the working face of the ring that is normally in contact with the bore.

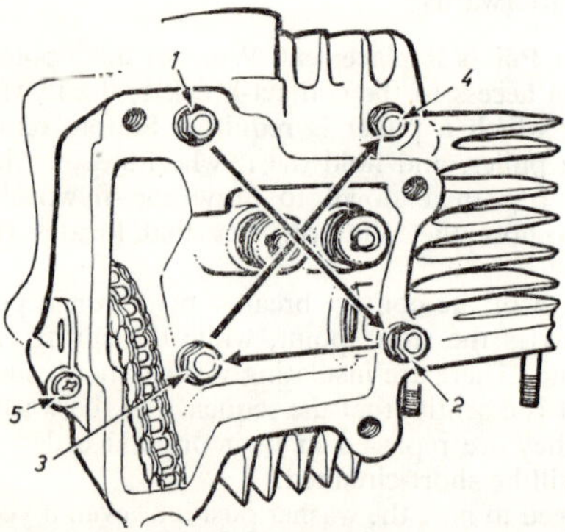

Fig. 30. Tightening head nuts, P.50 and PC.50. To avoid head distortion, follow this sequence when tightening the securing nuts

Next, carefully scrape all carbon from the grooves in the piston. Then roll each ring in turn round its groove—back to back—to check that it is free. Test the end gap by inserting each ring into the mouth of the barrel and use the piston, from the other end, to set it squarely about half an inch from the mouth. Measure the gap with a feeler. The gap must not exceed 0·024 in. If it does, the ring must be renewed. Honda rings come ready-gapped.

When replacing the rings on the piston, the manufacturer's mark must face upwards. Ease the second ring on first, working from the skirt end of the piston. Then add the oil ring—again from the skirt end—before finishing off with the top ring, which is fitted from the crown end.

Refit the piston—the arrow stamped on the crown must point downwards—and ensure that the circlips are fully home. Then rebuild the unit.

Further Dismantling. Further stripping of the P.50 unit is not recommended, and if it is contemplated reference should be made to the official Honda Workshop Manual for the full procedure.

In outline, the stripping procedure is—

1. Remove the engine and wheel unit.
2. Detach the engine/transmission unit from the wheel.
3. Disconnect the rear brake arm and remove the cam.
4. Remove the right-hand crankcase cover.
5. Remove the final driven sprocket and chain.
6. Remove the secondary driven and drive sprockets and chain.
7. Flatten the tab washer on the 14 mm clutch lock nut. Undo the nut and remove the primary driven and drive sprockets.
8. Remove the clutch.

On the engine itself, the cam chain tensioner and crankshaft are reached through the left-hand crankcase cover, after removal of the generator and back plate.

The head and cylinder must be detached as already described. Then—

1. Remove the reduction unit, as described earlier.
2. Remove the free pawl slider from the extreme upper rear of the crankcase.
3. Separate the left and right crankcase halves.
4. Withdraw the crankshaft from the right-hand case.

The tensioner in the left crankcase half should not be disturbed. If it requires renewal, the rivet on which it is fixed must have its head filed off and a new rivet must be used on re-assembly and properly peened over.

Note also, on re-assembly, that new gaskets (genuine Honda spares naturally) should always be used—and ensure that the two hollow dowels are correctly located in the left-hand crankcase half before this is refitted. The free pawl slider is replaced after the crankshaft has been re-installed.

OVERHAULING THE PC.50 POWER UNIT

So far as all work apart from complete stripping is concerned, the PC.50 unit is worked on as described for the P.50, save that for operations up to and including removal of the barrel, piston and rings it is not necessary to detach the engine from the frame. For further overhaul work, it may be found more convenient to do so, although much of the stripping can in fact be done with the engine still in place. Obviously, if it is intended to split the crankcase, removal is essential.

Removing the PC.50 Unit. Remove the carburettor cover and the H.T. lead. Then disconnect the decompressor cable at the engine end. Loosen the rubber induction pipe clamp and detach the carburettor. Take off the exhaust pipe, the chain case, and the pedal cranks. Disconnect the drive chain at the spring link.

Block up under the engine to support it. Then remove the nuts from the engine securing bolts, and ease the bolts away from the left-hand side. The electrical leads should now be unclipped, and the engine can be lowered from the frame.

Stripping the PC.50 Unit. Again, complete stripping is not advised. If you wish to do so, you should obtain the official Honda Workshop Manual. The general procedure, however, is—

1. Having removed the engine and drained the oil, detach the right-hand crankcase cover.
2. Remove the cylinder head and barrel. Also take off the piston.
3. Using Special Tool No. 07086–00101 take off the lock nut and detach the clutch assembly.
4. Remove the 17 mm circlip on the primary driven gear shaft and pull off the gear.
5. Detach the oil guide.
6. Remove the generator cover and detach the flywheel.
7. Remove the two bolts that hold the final drive sprocket on the left-hand side of the engine. Take off the sprocket.
8. Remove the kickstarter ratchet friction spring.
9. Remove the right-hand crankcase half after undoing its seven 6 mm screws.

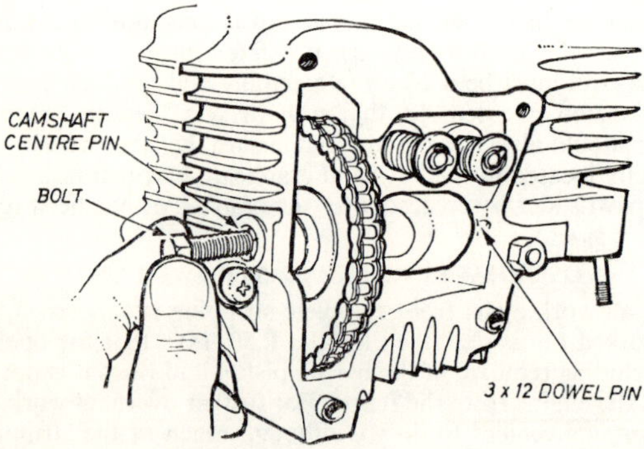

Fig. 31. Camshaft, P.50 and PC.50. How a bolt is used to extract the centre pin for camshaft removal

10. Remove the pedal shaft assembly.
11. Remove the main- and counter-shaft assemblies.
12. Loosen the 10 mm nut securing the starter's idler gear shaft and detach the gear.
13. Remove the crankshaft.

As with the P.50 engine, the camshaft chain tensioner is rivetted into place and need not be disturbed unless it is actually damaged.

Resetting the Valve Timing: PC.50. After the head has been removed on these o.h.c. machines, the valve timing must be reset. This is done by, first, setting the "T" mark on the flywheel against the line scribed on the crankcase, just forward of the point at which the wiring emerges.

With this position held, align the arrow on the camshaft sprocket so that it is pointing straight forward and fit the camshaft drive chain. The remaining procedure is as advised for the P.50.

OVERHAULING THE C.50 POWER UNIT

It is easiest to drain the oil before the engine is removed from the frame. Do this with the unit warm, so that the oil flows freely, and make sure that the oil filler cap is removed so that all lubricant is cleared.

Removing the Engine. Remove the air cleaner and detach the front shield. Then take off the silencer and exhaust pipe as a unit, and unbolt the foot-rest cross-bar. Take off the tool box, and detach the two 6 mm nuts that hold the carburettor. Unscrew the carburettor mixing chamber top and withdraw the slide, taping it to the frame so that it will not be damaged. Free the fuel pipe, and lift off the carburettor.

Remove the kickstarter and gearchange pedals, and then take off the left-hand crankcase cover. Break the electrical leads at the connectors, and turn the rear wheel until the spring link on the chain just clears the lower teeth of the final drive sprocket. Have a piece of soft wire handy. Remove the spring link, and break the chain. Don't let go of the ends, though. Instead, let the chain come clear of the sprocket and then wire the ends together. This will prevent it coiling up inaccessibly inside the chain case.

Pull the H.T. lead off the plug and remove the clip that locates the lead on the right-hand side of the crankcase. Free the brake pedal spring and the stop lamp spring.

Now place a support—a stout box, for instance—under the engine and take the 8 mm nuts off the engine mounting bolts. Support the unit with one hand and pull the bolts out. You can then lower the unit on to the support before taking it to the bench for stripping.

Engine removal is, of course, not necessary if all you plan is a top overhaul. Preparatory work where only the head is to be removed is confined to detaching the air filter and front shield; the carburettor; the exhaust system; and the H.T. lead.

Detaching the Cylinder Head. Whether the engine is on the bench or in the frame, this sequence must be followed if the head is to be lifted.

1. Remove the flywheel and the stator from the left-hand side of the engine.

2. Take off the two domed nuts and the two hexagonal nuts securing the cylinder head cover.
3. Detach the left and right-hand cylinder head side covers.
4. Rotate the crankshaft until the Woodruff key on the shaft and the O-mark on the camshaft sprocket are both on the centre-line of the engine.
5. Undo the three 5 mm bolts on the camshaft and the bolt between the head and the cylinder on the camshaft chain tunnel and carefully lift the head. Keep the cam chain tensioned as you do so.

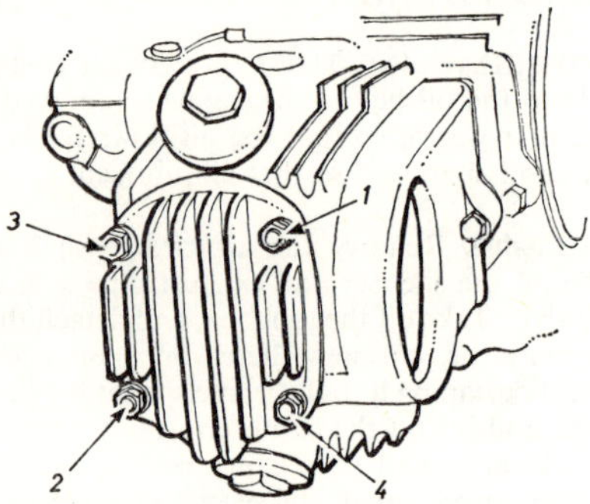

Fig. 32. Tightening head nuts, C.50. This sequence must be followed on the C.50 model to obviate distortion of the light alloy castings

Removing the Valves. The camshaft runs in machined housings, one at each side of the head. It will not normally have to be disturbed. To remove the valves, however, the rockers must first be removed. Each rocker is carried on a shaft that is a push-fit into its housings in the head. These shafts are accessible once the right-hand side cover has been removed from the head, simply pulling out. Each valve spring is then compressed with a compressor tool, in the normal way, until the valve cotters can be freed from their caps.

Removing the Cylinder Barrel. Undo the 6 mm cylinder flange bolts and take out the single bolt on the camshaft tunnel that holds the guide roller for the camshaft chain. The cylinder can then be drawn off its studs, supporting the piston as you do so.

Sealing O-rings. Note that rubber O-rings are used to seal the lubrication tunnels at the various joints. These O-rings *must* be refitted when the unit is assembled, or serious damage will be caused. Use new rings if the old ones are damaged. It is important to use genuine Honda replacements.

Removing the Piston. The gudgeon pin holding the piston to the connecting rod is located by circlips at each side. Free both of these, using circlip pliers, and press the pin sideways until the piston can be detached. An arrow mark, that must point to the front, is stamped on the piston crown to ensure that it is assembled the right way round. If you are at all unsure, note the direction of this mark *before* removing the piston, or lightly scratch your own mark on the crown (*not* on the piston walls) first.

The three piston rings are sprung out of their grooves in the normal way for cleaning and/or replacement.

Reassembly from Top Overhaul Stage. This degree of stripping is probably as much as most owners will attempt, and reassembly is mainly a reversal of the procedure outlined so far. Note, however, that if there is any reason to suppose that the cylinder head face is not absolutely true—e.g. signs of oil or gas leakage—it must be trued up on a face-plate before the engine is rebuilt. This work should be entrusted to a garage. If any dowel pins are removed, they should be refitted before assembly.

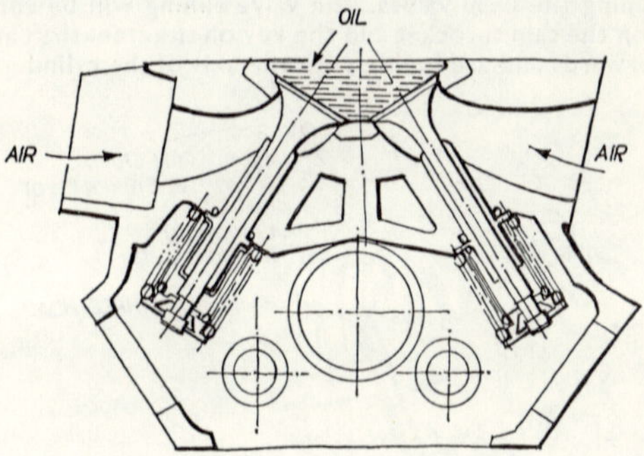

Fig. 33. How to check valve seats. After grinding-in the valves, refit them—together with their springs. Then invert the head and fill the combustion chamber with oil, using an old sparking plug in the plug hole. Block each port in turn with a large cork drilled to take the connector of a pump, and pressurize it. If any bubbles appear in the oil, the valve seat is not gas-tight and further grinding is required

Use a new cylinder base gasket, and lightly lubricate the bore before the barrel is slipped over the piston. Thread the camshaft drive chain through the tunnel, and then refit the guide roller. Build up the head complete with valve gear. Place the cam sprocket into mesh with the chain, and keep it under tension as the head is lowered towards the barrel, feeding the assembly through the tunnel and into the head. Note that the crankshaft key and the O-mark on the camshaft sprocket must be set in the same position as for dismantling—uppermost, and coincident with the engine centre line.

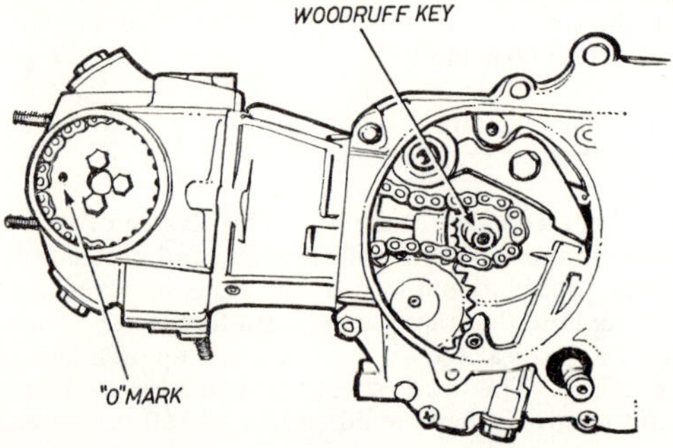

Fig. 34. Retiming the C.50 valves. The valve timing will be correct when the O-mark on the cam sprocket and the key on the crankshaft are pointing forwards and are in line with the axis of the cylinder

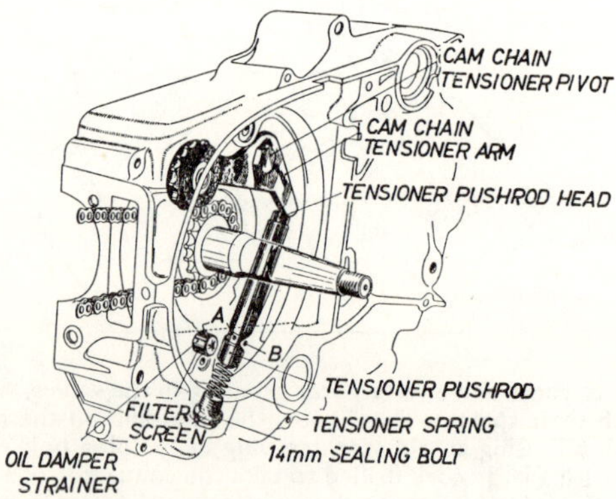

Fig. 35. Cam chain tensioner, C.50. Of oil-damped design, this tensioner should not normally require dismantling. If it is disturbed, note the position of the oil holes A and B

With the head home, lightly screw up the cam sprocket bolts and then install the cylinder head bolts. Recheck the valve timing—providing the relative position of the components is maintained as shown in Fig. 34 it *must* be correct—and tighten the camshaft bolts fully. Refit the cylinder head cover and tighten the nuts, evenly, a few threads at a time. Work from the left-hand upper to the right-hand lower; then from the right-hand upper to the left-hand lower, viewed from the *rear* of the engine. Finally, reset the ignition timing (as described for the P.50), and the tappet clearances.

Cam Chain Tensioner. Unless it is suspected that there has been blockage of the oil orifices in the damper, or that the spring has weakened, the cam chain tensioner in the left-hand crankcase will not normally need to be disturbed. It may, however, be dismantled—once the flywheel and stator have been removed—by undoing the 14 mm plug below the crankcase half and withdrawing the spring, push rod, and tensioner head.

Oil Pump Removal. A trochoid-type pump employing inner and outer rotors is used on the C.50. It is located behind the clutch, which must be removed before access can be gained. The pump itself is held by three 6 mm bolts and is backed by a gasket, which should be renewed on assembly. It is advisable to leave checking of the oil pump to a qualified Honda agent, since very fine clearances must be maintained if it is to work efficiently.

Clutch Removal. Two special tools—a holder for the clutch outer, and a T-spanner with which to undo the centre nut—are essential for use on the clutch, since the centre nut is of special castellated design. These should be purchased before you start work.

Before removing the right-hand crankcase cover, with the engine in the frame, drain the oil. Thereafter, the procedure is the same whether the unit is still in the machine or is on the bench.

Detach the kickstarter, and remove the 8 mm locking nut from the clutch adjusting bolt. Undo the cross-headed securing screws, and remove the crankcase cover. Undo the countersunk cross-headed screws holding the clutch outer cover and lift it away. Then lock the clutch with the holder tool, and undo the centre nut after flattening the lock washer. The complete clutch can then be pulled away from the shaft.

The plates and springs are held into the clutch body by a single large-diameter circlip let into a groove in the inner periphery of the drum. Use circlip pliers to compress this and lift it away. The plates and springs can then be detached. Note the order in which they are positioned as you remove them, and ensure that the same order is observed when the clutch is rebuilt.

Removing Crankcase and Transmission. This work involves splitting the crankcase, and is probably more than the average Honda C.50 owner would wish to tackle. For the benefit of experienced riders, however, I have included this sequence guide to complete dismantling.

1. Remove the engine unit from the frame.
2. Take off the final drive sprocket.

3. Remove the clutch.
4. Take out the gearchange spindle.
5. Take off the flywheel, stator, cylinder head, and barrel.
6. Remove the oil pump, and the primary drive gear wheel.
7. Remove the crankcase screws.
8. Place the unit on its left side and lift off the right crankcase half. This leaves the crankshaft and transmission in place in the left side of the case.
9. Lift out the crankshaft.
10. Remove the kickstarter spindle.
11. Lift out the main shaft, secondary shaft, and gear selection drum.

Carefully note the relative positions of all components before they are detached, and reassemble them in the same order after servicing.

One final word of warning. Honda units are designed to give an unusually high performance for their size, and of necessity the clearances and settings are to fine limits. Unless you are *sure* that you know exactly what you are doing, don't try to pull them to pieces. And even if you do, think twice first. It takes some thousands of miles for an engine unit to bed down and give of its best. Strip it unnecessarily, and you undo much of the good work that you have put in by way of careful maintenance and intelligent road use.

11 Suspension and brakes

THERE is a considerable family resemblance between the suspension systems on all 50 c.c. Hondas. Save for the P.50, with its rigid rear end, all have swinging fork suspensions coupled with swinging-link front forks. The sequence given here for servicing the C.50 can therefore be applied to all the 50 c.c. scooterettes and motor-cycles.

Leading Link Fork. To remove the spring units, detach the front wheel. This entails disconnecting the brake cable and the speedo drive, removing the wheel spindle and (where appropriate) the brake torque arm, and raising the machine slightly to allow the wheel and brake to drop out.

The upper end of the spring assembly is held to an abutment on the fork leg by a 7 mm nut locked by a pin. Withdraw the pin, undo the nut and lift out the cap washer and the rubber washer. On earlier models, the upper anchorage was by a pivot bolt, which is simply undone and removed to free the top mounting.

Next, release the leading link pivot bolt and withdraw it. This frees the complete spring/damper/link unit, which can then be pulled from the fork leg. The spring unit and damper is held to the link by a single pivot bolt (8 mm nut) and can be removed simply by undoing this. Take care not to drop the dust cap, dust seal and distance collar from the arm as the parts are separated.

The damper is a sealed unit and cannot be serviced. If it is faulty it will have to be replaced, but the spring itself can be detached. On the C.100, C.102, C.110 and C.115 type, hold the top collar in a vice and compress the unit sufficiently for the lock washer to be slackened. Then unscrew the collar, remove it and its buffer and anti-rattle sleeve, and detach the spring.

On the C.50 the operation is similar, save that the "collar" in this case is the threaded nut on which the spring bears.

The forks on the P.50 and PC.50 machines are virtually miniature versions of the C.50 fork and the method of dismantling is similar. After removing the wheel, release the 6 mm nuts on the top of the spring unit housings on the fork legs and the 8 mm nuts on the leading link pivot bolts. Spring away the mudguard stays that also locate on these bolts, and press out the pivots. The link and spring unit can then be pulled away from each fork leg. In this case there is no damper, the coil spring simply being threaded on to its upper and lower mounting collars, the upper one incorporating a rubber bump stop. The spring unit is held to the link by a pivot bolt locked by a 6 mm nut, and removing this frees it.

Steering Head Bearings. Here, again, the general layout is similar on all the Honda "50s," dismantling varying only in detail. The method described here for the C.50 can therefore be applied to the earlier models, and with a few reservations it also holds good for the mopeds.

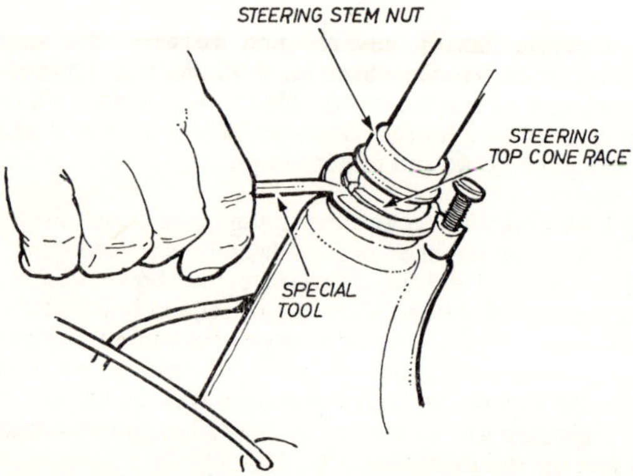

Fig. 36. Steering head, P.50. How the special tool is applied to the top race locking ring when adjusting the head bearings

To service the head bearings the forks must be removed. Detach the headlamp and disconnect the electrical leads at the connectors. Uncouple the speedometer drive and the front brake cable, disconnect the throttle cable at the carburettor, and detach the handlebars by removing the nuts from its securing bolts. Remove the front wheel, and take off the headlamp cowl. You can now reach the steering head.

Undo the two bolts that hold the steering head bridge to the fork legs, and unscrew the stem lock nut. Keeping the forks supported as you do so, use a C-spanner to undo the upper cone. As it comes clear of the lower race you will see the balls of the upper head bearing. Be careful not to lose any of these. It pays to have a tin handy and to place them in it as soon as you can reach them easily.

With the upper cone off, the fork can be slipped out of the steering head. Again, take care not to lose any of the balls from the lower bearing. And don't put them in the same tin as the upper ones. If they are to be re-used, these bearing balls must go back on their original tracks.

The tracks, cones and balls must be carefully examined for wear. If there is noticeable pitting or cracking renew the complete bearing. The races that carry the tracks are an interference fit in the head and can be driven out from the far side using a long soft-metal drift and a hammer. Seat the new ones by placing them against their housings and tapping them into place with a block of wood.

Where the original parts are to be re-used, clean them with petrol and allow them to dry. Then liberally grease the tracks and position the balls in them. The grease will hold them there while the fork is offered up and the top cone tightened.

Note that the accuracy of the steering will depend upon the care with which this work is done. If the cone is too tight the steering will be stiff and the machine will be uncontrollable. If it is too loose, there will be fore-and-aft movement of the forks and the machine will tend to wander.

A literally rule-of-thumb guide that I have found to work well is to set the cone so that the forks will flick easily from side to side under a gentle push, but with no trace of movement when one thumb is placed on the top cone/steering head joint and the forks are rocked backwards and forwards.

Fork Removal, P.50 and PC.50. These additional points should be noted on the mopeds, in which the handlebars are held by an expander plug in the stem. While the foregoing working instructions apply in general, the sequence is to remove the front wheel first, followed by the headlamp and electrical leads, the horn and front carrier. Then the expander plug set bolt in the centre of the bars is loosened by a couple of turns or so and given a sharp tap with a soft-faced mallet or by a hammer with a block of wood interposed. This drives the expander down the stem and frees the bars, which can then be pulled up and out.

When re-assembling, note that the height of the bars can be varied and set them as you require. Do this by lightly turning the expander bolt so that some movement is still possible and obtain the desired height before rocking it. Make sure, too, that the bars are square to the forks first.

Rear Suspension. Here, again, it is easy to see the family resemblance in Honda rear springing layouts and the C.50 method of fork removal and bearing replacement can be taken as typical.

Having removed the rear wheel, detach the chain case, the chain and—where applicable—the final drive flange. Then remove the nut on the main pivot bolt and detach the upper and lower nuts on the spring unit pivots. Take off the units, withdraw the pivot bolt, and the rear fork can be lifted off. If the pivot bushings in the fork are worn they can be driven out from the rear and new ones pressed into place.

BRAKES

Routine maintenance on the braking system amounts only to adjustment which, on all Hondas, is effected simply by mean of screw-types adjusters on the operating cables or rods. The wheel on which you are working should be raised, and the adjuster turned until the brake is just binding. It should then be released, half a turn at a time, until the wheel is again able to rotate with no trace of a shoe rubbing on the drum.

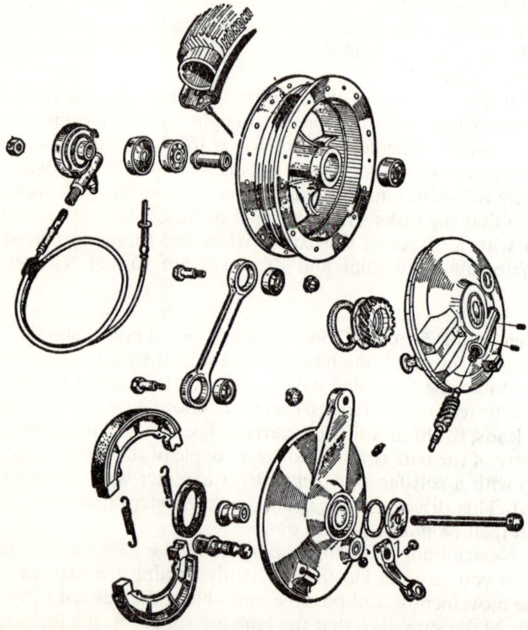

Fig. 37. Front brake and hub, C.100 and C.102. On early versions of the Honda a backplate secured by a torque arm was used. Later models have the slotted type of backplate. Otherwise the front brakes of all models are of similar design. Both types are shown here

Where there is no further adjustment left—by which time the brake operating arm will form more than a right angle with the cable when the brake is applied—re-lined shoes must be fitted. Don't try to re-line the old ones yourself—it is rarely possible to get just the right radius to match the drum—but use genuine Honda shoes instead.

The work involved is extremely simple. Remove the wheel and take off the brake plate, complete with shoes. Clean all dust out of the drums. Take the back plate to your bench, and you will note that the shoes locate on a fixed pivot at one end and against a cam on the operating arm at the other. They are held together by springs—tension springs in the case of the scooterettes and motor-cycles; a single horse-shoe spring on the mopeds.

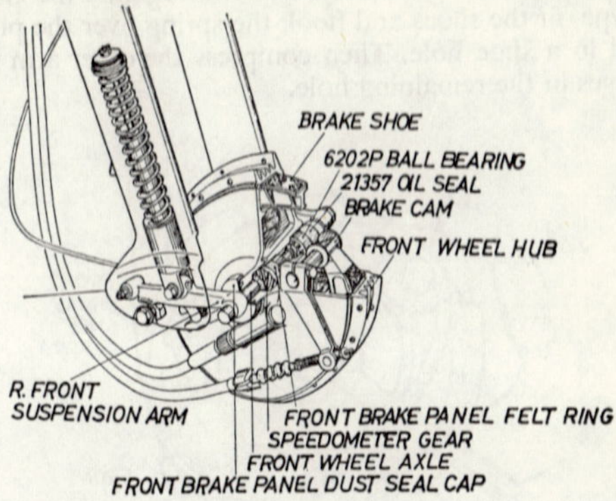

Fig. 38. The C.50 front suspension and brake layout

Make a careful note of the way the springs are fitted. With the horse-shoe type, use a screwdriver to lever the spring out of the shoe flange holes and the shoes will then lift away. With tension springs, pull the shoes apart until the ends can be freed from the cam and the pivot. When the

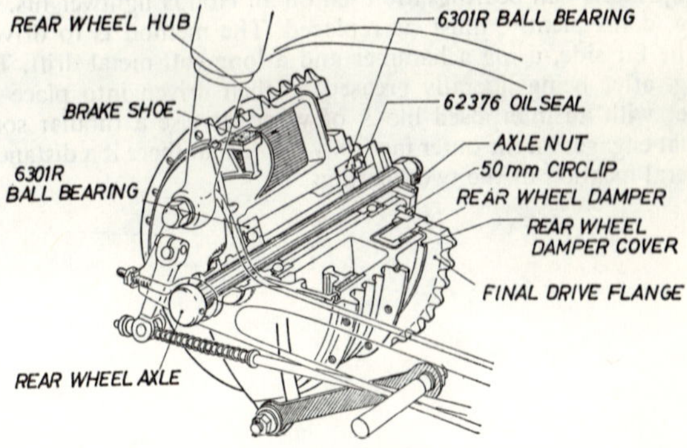

Fig. 39. The C.50 rear brake and hub

shoes are off, disconnect the springs from the flanges. Thoroughly clean the backplate and check the operating arm. It must move freely.

Fit the new shoes by linking them with the springs, and expanding them outwards so that one end can be engaged with the pivot. Then pull them apart once more to seat the other ends against the cam. With the horseshoe type, fit the shoes and hook the spring over the pivot with one end engaged in a shoe hole. Then compress the other arm of the spring until it engages in the remaining hole.

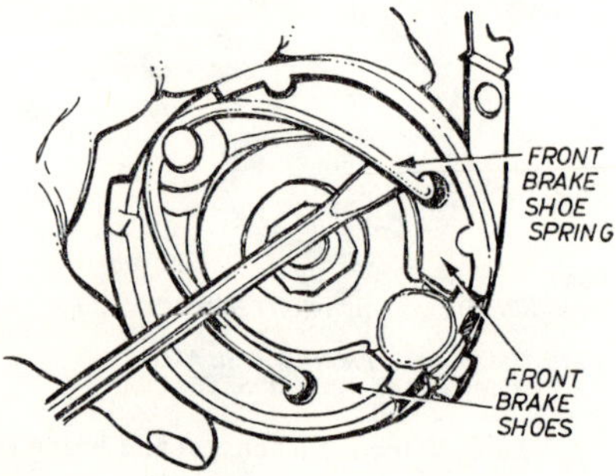

Fig. 40. Horseshoe-type springs, P.50 brakes. The P.50 range is unusual amongst Hondas in using this type of pull-off spring for the brakes. It is levered out by use of a screwdriver, as shown here

WHEEL BEARINGS

Non-adjustable ball bearings are used on all Honda lightweights. A worn bearing, consequently, must be replaced. The method is to drive it out from the far side, using a hammer and a long soft-metal drift. The new bearing, after being liberally greased, is then driven into place using a hammer with an interposed block of wood or else a tubular soft-metal drift that engages on the outer face only. Note that there is a distance collar for lateral location of the two bearings.

Appendixes

Appendix A
Facts and figures

Data	C.100	C.102	C.110	C.114
Bore	40 mm	40 mm	40 mm	40 mm
Stroke	39 mm	39 mm	39 mm	39 mm
Capacity	49 c.c.	49 c.c.	49 c.c.	49 c.c.
C.R.	8·5:1	8·5:1	9·5:1	9·5:1
Output	4·5 b.h.p. at 9,500 r.p.m.	4·5 b.h.p. at 9,500 r.p.m.	5 b.h.p. at 9,500 r.p.m.	5 b.h.p. at 9,500 r.p.m.
Ignition timing (full advance)	35° b.t.d.c.	5° b.t.d.c. (Stable)	35° b.t.d.c.	35° b.t.d.c.
Ring gap (compression)	0·003/0·010 in.	0·003/0·010 in.	0·003/0·010 in.	0·003/0·010 in.
Ring gap (oil control rings)	0·003/0·010 in.	0·003/0·010 in.	0·003/0·010 in.	0·003/0·010 in.
Exhaust valve length	60·6–60·8 mm (2·3858–2·3937 in.)			
Inlet valve length	61·1–61·3 mm (2·4055–2·4134 in.)			
Valve spring free length (outer)	27 mm (1·063 in.)		28·4 mm (1·118 in.)	
Valve spring free length (inner)	27·8 mm (1·0945 in.)		26·9 mm (1·0591 in.)	
Pushrod length (inlet)	187·4 mm (7·3779 in.)			
Pushrod length (exhaust)	170·5 mm (6·7126 in.)			
Contact-breaker gap	0·014 in.	0·014 in.	0·014 in.	0·014 in.
Sparking plug gap	0·024 in.	0·024 in.	0·024 in.	0·024 in.
Tappet clearances	0·002–0·004 in. cold			
Main jet	88/95	88/95	85/88	85/88
Slow running jets	35	35	35	35
Throttle slide	2	2	2 (Pw 16)	2 (Pw 16)
Jet needle	13302	13302	16302	16302
Turns out, air screw adjuster	1/1¼	1/1¼	1/1¼	1/1¼
Oil capacity	1⅛ pints	1⅛ pints	1⅛ pints	1⅛ pints
Oil grade (winter)	20 S.A.E.	20 S.A.E.	20 S.A.E.	20 S.A.E.
(summer)	30 S.A.E.	30 S.A.E.	30 S.A.E.	30 S.A.E.
Petrol tank capacity	6·4 pints	6·4 pints	1·6 gal.	1·6 gal.
Sparking plug	Type N.G.K. C7 HW (10 mm × 12·7 mm reach)			
Typre pressure:				
front	22 p.s.i.	22 p.s.i.	22 p.s.i.	22 p.s.i.
rear	28 p.s.i.	28 p.s.i.	28 p.s.i.	28 p.s.i.
rear (pillion)	32 p.s.i.	32 p.s.i.	32 p.s.i.	32 p.s.i.
Bulbs:				
head	6 V 15/15 W	6 V 20/20 W	6 V 15/15 W	6 V 15/15 W
tail	6 V 1·8	6 V 1·8 W	6 V 1·8 W	6 V 1·8 W
stop	6 V 6 W	6 V 6 W	6 V 6 W	6 V 6 W
winker	6 V 8 W	6 V 8 W	6 V 8 W	6 V 8 W
neutral	6 V 3 W	6 V 3 W	6 V 3 W	6 V 3 W
	small bayonet	small bayonet	small bayonet	small bayonet
winker indicator	—	—	"	"
speedo	6 V 1·5 W	6 V 1·5 W	6 V 1·5 W	6 V 1·5 W
Weight (dry)	143 lb.	154 lb.	145 lb.	146 lb.
Length	71·25 in.	71·25 in.	67 in.	67 in.
Width	22·25 in.	22·25 in.	22·24 in.	22·24 in.
Height	37·25 in.	37·25 in.	36·25 in.	36·25 in.
Ground clearance	5½ in.	5½ in.	5·9 in.	5·9 in.
Tyre sizes	2·25 × 17 in. front and rear			

C.50	P.50	PC.50	PC.50 o.h.v.	PF.50
39 mm	42 mm	42 mm	42 mm	42 mm
41·4 mm	35·6 mm	35·6 mm	35·6	35·6 mm
49 c.c.	49 c.c.	49 c.c.	49 c.c.	49 c.c.
8·8:1				
4·8 b.h.p. at	1·75 b.h.p. at	1·75 b.h.p. at		
10,000 r.p.m.	5,750 r.p.m.	5,750 r.p.m.		
		Not quoted		
0·004/0·0118 in.	0·004/0·012 in.	0·004/0·012 in.	0·004/0·012 in.	0·004/0·012 in.
0·012 in.				
65·3 mm	49·5 mm	49·05 mm	47 mm	47 mm
(2·573 in.)	(1·945 in.)	(1·928 in.)	(1·85 in.)	(1·85 in.)
66 mm (2·60 in.)				
28·1 mm	19·4 mm	22·92 mm	24·5 mm	24·5 mm
(1·11 in.)	(0·764 in.)	(0·902 in.)	(0·965 in.)	(0·965 in.)
Not quoted				
		Not applicable	Not quoted	
		Not applicable		
0·012–0·016 in.	0·012–0·016 in.	0·012–0·016 in.	Not applicable	
0·024–0·028 in.	0·024–0·028 in.	0·024–0·028 in.	0·024–0·028 in.	0·024–0·028 in.
0·002 in. cold	0·002 in. cold	0·002 in. cold	0·002 in. cold	0·002 in. cold
70	58	55	62	62
35	35	35/45	35	35
2	2·5	2	2·5	2·5
			5° 30′, 2·03 mm	2·1 + 2·8 P–R,
				OD 4·2
			As required	
13239	2° 30′, 2 mm	015301–3		
1⅛/1½	1/1⅛	1⅛/1⅜		
1·4 pints	1·2 pints	1·4 pints	1·3 pints	1·3 pints
20 S.A.E.	10W/30	10W/30	10W/30	10W/30
30 S.A.E.				
6·7 pints	4·8 pints	5·6 pints	5·6 pints	4·1 pints
N.G.K. C7 HS	N.G.K. C6 HB	N.G.K. C6 HB	N.G.K. C7 HS	N.G.K. C7 HS
24 p.s.i.	19 p.s.i.	28 p.s.i.	18·5 p.s.i.	26 p.s.i.
30 p.s.i.	27 p.s.i.	28 p.s.i.	25·6 p.s.i.	28 p.s.i.
34 p.s.i.	—	—	—	—
6 V 15/15 W	6 V 10 W	6 V 10 W	6 V 10 W	6 V 10 W
6 V 5 W	6 V 5 W	6 V 5 W	6 V 5 W	6 V 5 W
6 V 18 W	6 V 5 W	6 V 5 W	6 V 5 W	6 V 5 W
6 V 18 W	6 V 18 W	6 V 18 W	Not applicable	
—	—	—	Not applicable	
—	—	—	Not applicable	
6 V 1·5 W	—	—	Not applicable	
152 lb.	99·1 lb.	105·5 lb.	110 lb.	98 lb.
70·7 in.	65·7 in.	69·2 in.	69·1 in.	98 in.
25·2 in.	24·4 in.	23·6 in.	23·6 in.	25 in.
38·4 in.	40·2 in.	40·0 in.	40·7 in.	41 in.
5·1 in.	4·3 in.	5·9 in.	5·5 in.	5 in.
2·00 × 17 front	2·00 × 19 front		2·00 × 19 front	2·00 × 17
2·25 × 17 rear	2·25 × 19 rear		2·25 × 19 rear	front and rear

Appendix B
Carburettors

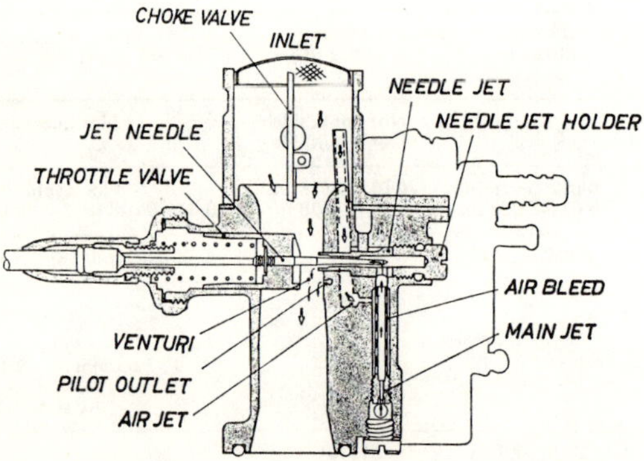

Fig. 41. The C.50 carburettor. Air and fuel flow in the instrument is shown in this diagrammatic view

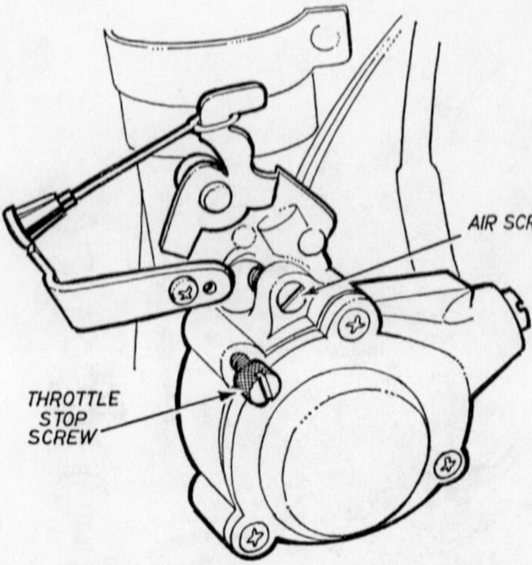

Fig. 42. Running adjustments, P.50. Common to all the machines, these two adjusters are perhaps not quite so obvious on this model

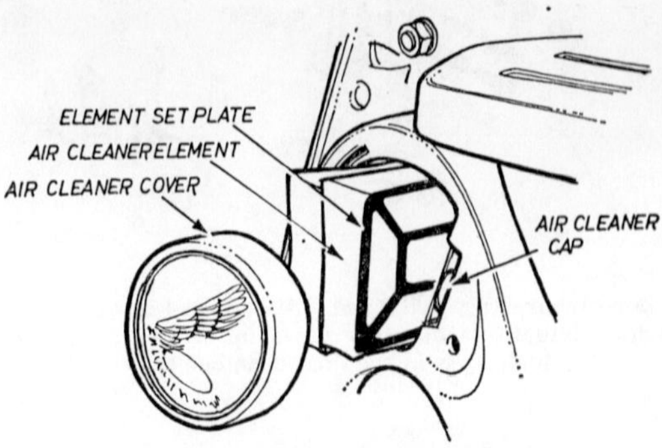

Fig. 43. Air cleaner element. Poor performance often proves to be due to a dirty air filter. This is the P.50 type. Others are similar in design, but are positioned differently

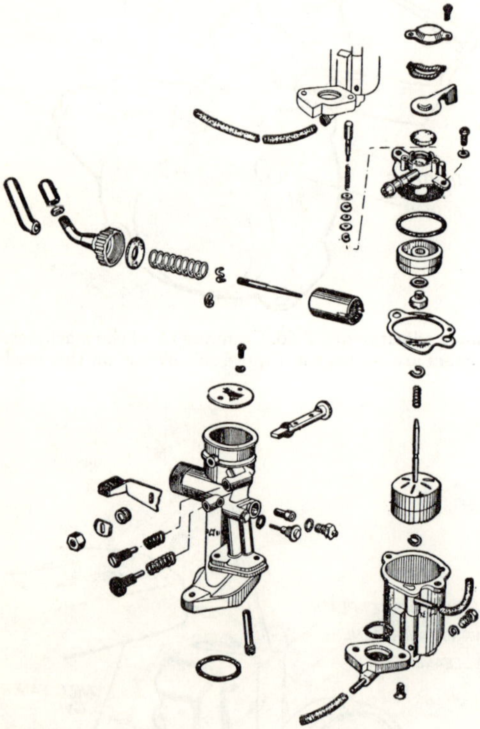

Fig. 44. Keihin carburettor, C.100 and C.102. Bolted direct to the cylinder head, this downdraught carburettor has a butterfly-type choke and the fuel tap is in the float chamber top

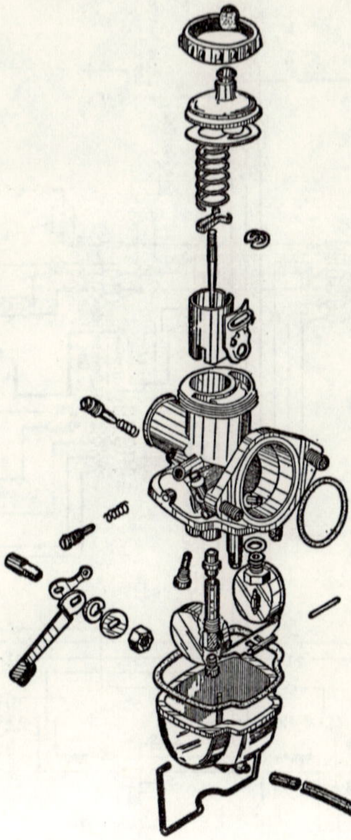

Fig. 45. Keihin carburettor, C.110 and C.114. On this carburettor the choke is of slide type and an unusual feature is that hot oil is circulated past the mixing chamber to prevent icing

Appendix C
Wiring diagrams

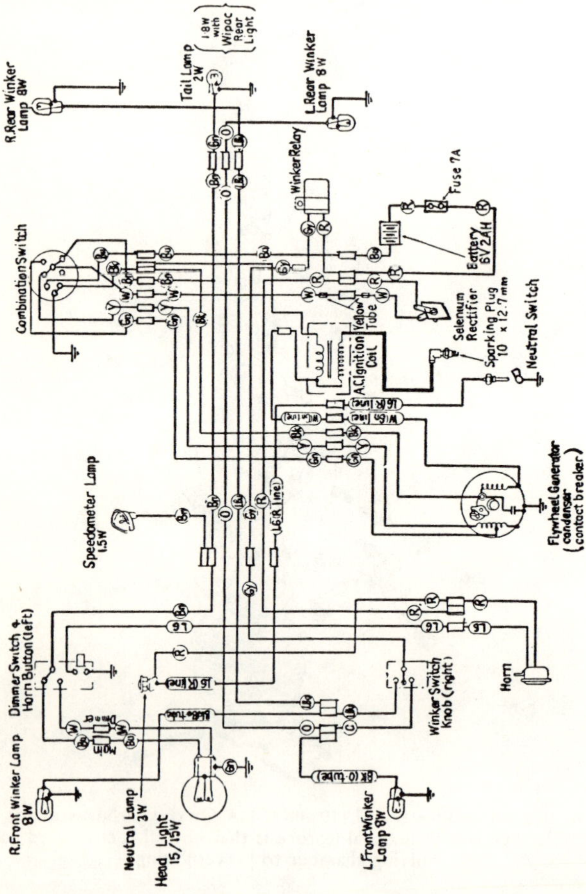

Fig. 46. Wiring diagram, Honda C.100. Bk = black. Bu = blue. Bn = brown. Gy = grey. Gn = green. L.Bu = light blue. LG = light green. O = orange. R = red. W = white. Y = yellow. Bu-tube = sheathed in light blue plastics. O-tube = sheathed in orange plastics. R-line = red line. Gn-line = green line

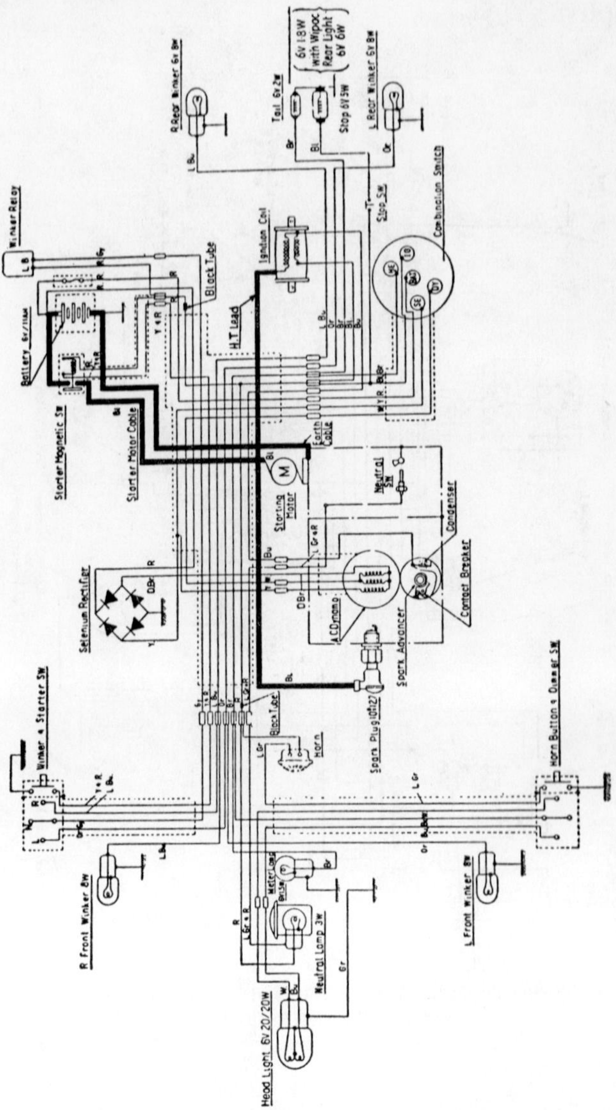

Fig. 47. Wiring diagram, Honda C.102. For abbreviations see Fig. 46—wiring diagram of Model C.100—which is similar, save for the use of a starter motor on the Model C.102

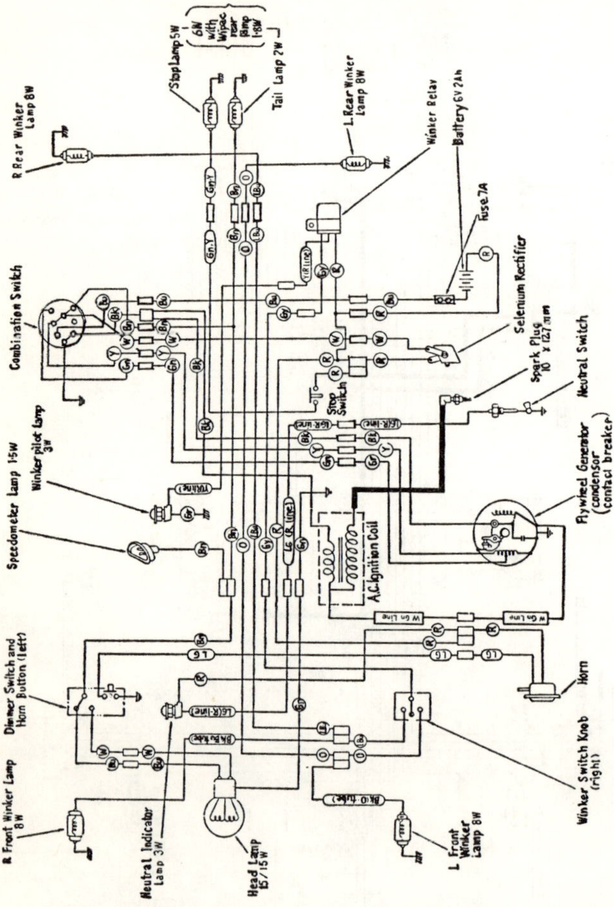

Fig. 48. Wiring diagram, Models C.110 and C.114. For abbreviations see Fig. 46—wiring diagram of the Model C.100

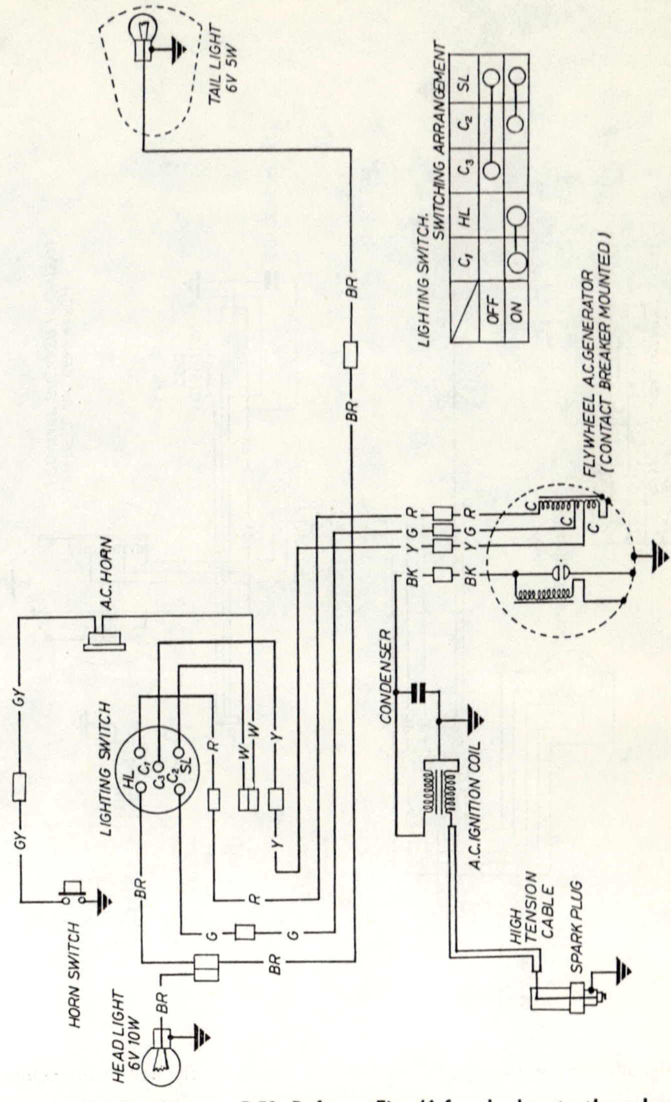

Fig. 49. Wiring diagram, P.50. Refer to Fig. 46 for the key to the colour coding of the wires

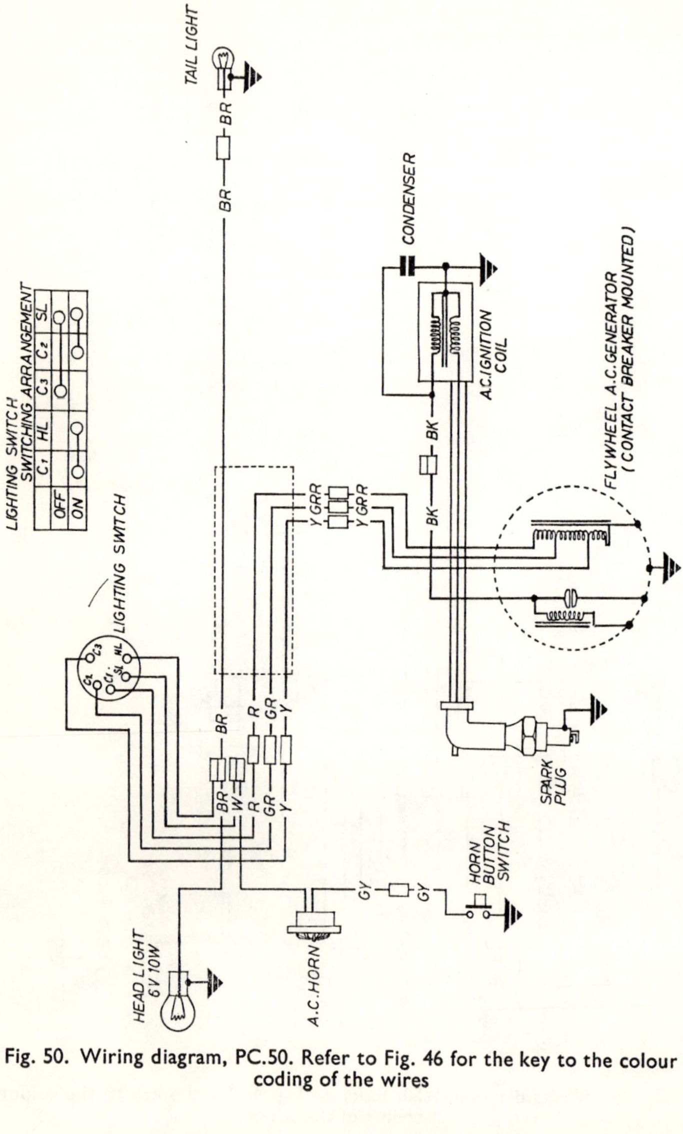

Fig. 50. Wiring diagram, PC.50. Refer to Fig. 46 for the key to the colour coding of the wires

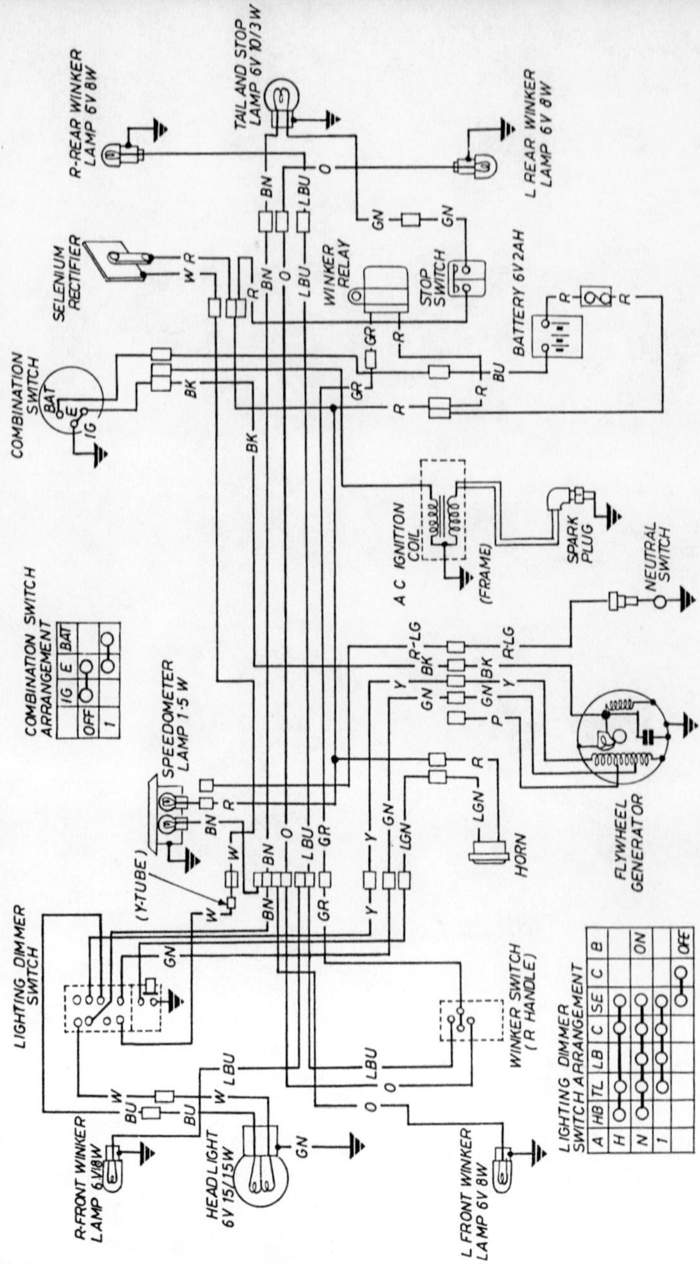

Fig. 51. Wiring diagram, C.50. Refer to Fig. 46 for the key to the colour coding of the wires

Fig. 51. Wiring diagram, C 50. Refer to Fig. 46 for the key to the colour coding of the wires

Index

AIR cleaner, 57, 99
filter, 20

BARREL removal, 62, 65
Basic cycle, 12–13
principles, 11–23
Battery, 17, 59
Bearings, 92
Big end, 12
Bottom dead centre (B.D.C.), 12–13
Brakes, 21–3, 87–90
adjustment, 58
faults of, 41–3
Braking, 9
Bulbs, 59

CAM chain tensioner, 84–5
Cams, 15
Camshaft removal, 70
Carburettor, 18–21, 28, 96–9
adjustment, 55–6
cleaning, 56–7
fuel taps, 54
throttle needle adjustment, 54
Chain adjustment, 59
faults in, 41
Changing down, 8
Clutch, 7–9, 20–1
adjustment, 52–3
automatic, 69
faults, 39
removal, 70, 85
Condenser, 18
Connecting rod, 12
Contact-breaker, 17, 74
gap, checking and adjusting, 49–50
points, setting, 77
replacement, 77
Cornering, 9–10
Crankcase and transmission removal, 85
Crank pin, 12–13
Cycle parts, 21
Cylinder head, 12; detaching, 71, 81

DECARBONIZING, 65, 75, 78
Decompressor cable adjustment, 53
Dismantling, complete, 68–70, 80
Drain plug, 45

EARTH, 15
Engine, 11
overhauls, o.h.c. models, 73–86
pushrod o.h.v. models, 61–72
reassembly, 71–3
removing the, 73
Exhaust stroke, 13

FACTS and figures, 96–7
Faults, tracing, 28–41
Float-chamber, 18
mechanism, 29
Fork removal, 89
Frame numbers, 4
Fuel system, 28–9
tap, 58

GEARS, 7–8
faults in, 39
Generator, 17

HANDLING, 5–10
Head removal, 62, 64–5, 79–80
Honda 50 c.c. range, 1–4
C.50, 1–3, 7–8, 15, 45, 50–2, 57, 81–6, 89, 91, 96–7, 105
C.100, 1, 6–8, 13, 18, 21, 51–3, 55, 57, 70, 87, 90, 98, 100
C.102, 1, 7–8, 13, 18, 21, 51–3, 55, 57, 68–9, 87, 90, 98, 100
C.110, 1–2, 6, 13, 24, 52, 55, 57–9, 68–70, 87, 99, 102

Honda 50 c.c. range (*contd.*)—
C.114, 1–2, 6, 13, 52, 55, 57–9, 63, 69–71, 99, 102
P.50, 2, 3, 7, 16, 21–2, 47, 50, 53–4, 56–7, 71–9, 87–9, 92, 103
PC.50, 3, 24, 47, 50, 53, 56–7, 71–81, 87, 89, 104
Hydraulic damper, 21

IDLING, 37
Ignition system, 15–18, 29
timing, 76
Inlet valve, 12

KICKSTARTING, 6

LEADING link fork, 87
Lighting faults, 32–4
Lubrication system, 15

MAGNETO, 18
Maintenance, 42–59
Misfiring, 38

NEEDLE jet, 19
Neutral gear, 8
Noises, 40

OIL change, 32, 44–5
pump removal, 85
Open-enders, 24–5
O-rings, sealing, 82
Overflooding, 29
Overhead camshaft layout, 14
valve layout, 13

PISTON, 12
fitting, 66
removal, 64, 66, 83
rings, renewing, 77
Power unit overhaul, 73–86

R.A.C./A.C.U. Learner Training Scheme, 5
Reassembly, 68, 83
Running faults, 36

SCREWDRIVERS, 24–6
Self-tuition, 6
Short circuit, 33
Silencers, 59
Small end, 12
Spanners, 24–6
Sparking plug, 17, 29–30; gap, checking and adjusting, 48–9
Speed control, 8
Springing faults, 41
Starting, 6, 12, 20
faults, 35
Steering faults, 40
head bearings, 88
Stoplight timing, 59
Stripping: C.100 and C.102, 61–2
C.110 and C.114, 62–71
Sump, 15
Suspension, 87–92
Swinging fork layout, 21

TAPPET block, 15
checking and adjusting, 46–8, 61
faults, 31, 42
Task system, 43–5
Throttle, 6–8, 18
valve assembly, 19
Timing, checking, 50
Tools, 24–7
Top dead centre (T.D.C.), 12–13
Training schemes, 5
Transmission faults, 32

VALVE gear, 63
grinding, 75
guides, 75
removal, 65, 74
retiming, 70, 81, 84
seating, 67
Venturi, 18–19

WHEEL removal, 58
Wiring, 30
diagrams, 33, 100–5